Biotechnology

From A to Z

Biotechnology
From A to Z

by
WILLIAM BAINS

Oxford   New York   Tokyo
OXFORD UNIVERSITY PRESS

Oxford University Press, Walton Street, Oxford OX2 6DP
Oxford   New York
Athens   Auckland   Bangkok   Bombay
Calcutta   Cape Town   Dar es Salaam   Delhi
Florence   Hong Kong   Istanbul   Karachi
Kuala Lumpur   Madras   Madrid   Melbourne
Mexico City   Nairobi   Paris   Singapore
Taipei   Tokyo   Toronto
and associated companies in
Berlin   Ibadan

Oxford is a trade mark of Oxford University Press

Published in the United States
by Oxford University Press Inc., New York

© William Bains, 1993

First published 1993
Reprinted 1993, 1994

A catalogue record for this book is available from the British Library

Library of Congress Cataloging in Publication Data
(Data available upon request)

ISBN 0 19 963334 7

Printed in Great Britain on acid-free paper by
Bookcraft (Bath) Ltd, Midsomer Norton, Avon

Introduction

by G. Kirk Raab
President and Chief Executive Officer
Genentech, Inc.

Right now the biotechnology industry stands on a threshold. It offers the public significant promise that can seem beyond belief. Yet the industry has reached a level of success and maturity where it is becoming, on some levels, mundane. From the cheese we eat that is made with bioengineered rennin to the news reports we hear of criminals convicted based on minute bodily evidence, biotechnology is already a part of our daily lives.

The innovation of biotechnology is harnessing nature's own biochemical tools, hewn for their tasks through years of evolution, to make possible new products and processes.

Biotechnology's greatest impact so far has been in health care. Large-molecule protein pharmaceuticals — now standard treatments for serious diseases — are the most visible contributions. Safe and abundant insulin for diabetics and growth hormone for those deficient in the protein offer thousands the promise of a healthy, normal life. Agents to promote the growth of needed blood cells for cancer-chemotherapy and kidney-dialysis patients greatly enhance these patients' quality of life. A fast-acting 'clot buster' for stopping heart attacks saves lives every day. Even before therapeutics are prescribed, biotechnology has given physicians the power to diagnose disease, or even the risk of disease, early, when medical care can bring the greatest benefit.

The advances, and the impact, will continue. Molecular biology is increasing our understanding of human health and disease at a spectacular rate. With the insight already gained, scientists can now design pharmaceuticals and therapeutic strategies to target specific diseases without the toxic side effects that typically accompany traditional pharmaceuticals. Many such pharmaceuticals have been identified and are currently being tested to treat life-threatening diseases (such as cancer), severe infections, and respiratory disorders like cystic fibrosis, chronic bronchitis, and asthma.

The health-care industry's response to the AIDS epidemic exemplifies a revolution in the drug discovery process. In the dozen years since AIDS has been identified, researchers have identified the virus that causes the disease, characterized it, and used the knowledge to design dozens of custom-built pharmaceuticals — many now undergoing clinical testing — in an attempt to treat or prevent the disease. This rate of discovery and development is unprecedented in medical history.

Bodily systems that have been frustratingly enigmatic are now being understood and utilized to treat human disease stemming from malfunctions in these systems. The immune system, the brain and nervous system, the intricate genetic system controlling cell growth and differentiation, for example, are all beginning to make sense. As scientists decipher the secrets of these systems, it will become increasingly clear that biotechnology's day in the clinic has only dawned.

But biotechnology is not just about health care. The ability to harness living systems is already providing solutions to some of society's woes. Biotechnological processes are inherently low-energy, renewable technologies that fit more comfortably with the ethos of the 1990s than energy- and waste-intensive solutions of the past. Crops genetically engineered to be less perishable and to resist disease are being produced and consumed, leading to reduced use of chemical pesticides and to a longer window for produce distribution, potentially to third-world countries. Micro-organisms are being used to clean up oil and chemical spills. A sensitive forensic technique called DNA fingerprinting is providing potent new tools to fight crime. Bioengineered, biodegradable plastics are offering early solutions to the world's growing waste-disposal problem. And enzymes are finding their unique powers of catalysis in demand in places as diverse as high-technology chemical plants and the laundry room at home.

In this decade, I think biotechnology will take a firm and giant step across the threshold on which it now stands. Ralph Nesbitt called the 1990s the decade of biology. Not because he expects fundamental new discoveries but because products of biotechnology will become integrated into everyday life to the extent that chemicals, computers, and traditional pharmaceuticals are now.

This means that more and more of us will become increasingly involved with biotechnology, as a science, as an industry, and as a supplier and consumer of marketable and consumable products.

The leap toward widespread public understanding and involvement will be difficult, but we are not starting from scratch. The public is already watching and listening to news reports about biotechnology progress and the emerging policy issues surrounding that progress.

Public interest in the regulation of biotechnology was most visible in the 70s and 80s. Opposition to perceived dangers of genetic manipulation

— dangers that have failed to materialize — made headline news. It sadly continues today as a group of US chefs choose to boycott genetically engineered tomatoes.

Since its inception, the industry has done its best to accede to public and legislative demands for openness and regulation, having learned the lesson of the nuclear industry, whose secrecy condemned it to almost reflex distrust by the public. It started in 1974 with the Berg letter. A group of scientists in the nascent field of recombinant DNA technology, led by biotechnology pioneer Dr Paul Berg (who later won a Nobel Prize), called for a moratorium on their own research so that the potential risks could be evaluated. The industry's desire to inform the public about itself has continued unabated ever since, to the extent that every Western country now has its own collection of industrial and scientific associations dedicated to spreading the word about biotechnology. The message is sometimes presented in terms that would baffle anyone but a scientist. But the spirit is willing even if the vocabulary is weak.

An equally visible aspect of public involvement has been financial, as investors tried to buy a ride on this wave of the future. The extraordinary enthusiasm on Wall Street at the initial offering of Genentech in 1980 has waxed and waned over the years, as the market will do, and reappeared full-force as biotechnology became the fail-safe hot tip in the early 1991 market — only to wane again and then settle into a cautiously watchful 1992 market. All along, I think there has been more at work than pure financial motives. People want to own a piece of this technology.

And people are increasingly demanding the products of biotechnology. Whether as a conscious decision to 'go green' with biodegradable plastic or a medical need for a biopharmaceutical — a need so urgent as to spur consumers to prod biopharmaceutical developers and government regulatory agencies to speed drug availability. Some products, however, still face tough public scrutiny, despite obvious benefits. A cow made plump with a genetically engineered version of a natural beef growth hormone is suspect in the eye of a public unclear on the technology involved.

Despite public involvement to date, knowledge of the basics of biotechnology is not what it should be, and certainly not what it will need to be. A recent survey in Europe showed that public understanding of biotechnology is patchy at best. Might a lack of public understanding in Europe contribute to the relatively hostile public environment for biotechnology research and manufacturing that exists there? Can governments make sensible decisions about application and regulation of this technology when more than half of the general public does not know that DNA has to do with life, and not geology or space science?

INTRODUCTION

As biotechnology steps into the room of the future, the public needs to be there to meet it — to accept it as a critical technology that can help us toward a brighter future as we face the intensifying challenges of disease, ageing, world hunger, and waste management.

Those who interpret and communicate the industry — the media, financial analysts, government regulatory agencies, educational institutions, and, of course, the industry itself — must all play a role in educating the public. To do so, and to do so accurately, they need to know what the new technology is about, and what it can and cannot do. The explanations of ideas and terms in biotechnology that this book comprises will help bring about that understanding. And so will help bring about the day when we cannot imagine life without biotechnology any more than we can imagine life without computers.

How to use this book

This is not a textbook. It is an extended glossary of terms which will give you a quick insight into a term or concept in one of the most exciting areas of applied science. The entries give a quick description of the concept, mention any related terms or ideas, and where relevant describe what practical application that idea has. It is a book for dipping into, for quick reference when puzzled or momentarily lost. You do not have to read half of it to get to the explanation you want.

The book is for the non-expert, and so it does not assume that you have a Ph.D. in biochemistry. Some familiarity with the basic ideas of modern biology is useful: if you are floored by terms such as 'bacterium' or 'DNA', then this book is definitely not for you. If you can get that far, then we can work together.

The book consists of around 280 entries for key terms or concepts in biotechnology. They are arranged alphabetically, so you can thumb through them to find what you want or just browse and see what catches your eye. However, if you want an answer fast then turn to the index at the back. Any term that is described in the book is listed there.

Explaining each term completely in a self-contained entry would mean a lot of redundancy, so many of the entries cut short discussions of topics which are covered elsewhere in the book, and simply refer you to that entry. You can pick your way through these connections until you feel that you have found enough of the background to the term you started with.

Acknowledgments

While the vagaries of what I have included in this book and how I have described it are entirely my fault, the errors and omissions would be much more numerous without the generous help of my colleagues. My particular thanks to Prof. Tony Atkinson and Dr Peter Hammond (Centre for Applied Microbiological Research), and Dr Andy Richards (Chiros Ltd.) for substantial contributions. I am also grateful for the helpful comments of Nick Ashley, Jane Devereaux, and Lyn Scott (PA), Dr Brian Richards (British Biotechnology), Dr Chris Lowe (Cambridge University), and the various referees at OUP, and, of course, to Kirk Raab for the Introduction. My thanks to Prof. Roger Whittenbury for encouraging words at critical times, and to my family for putting up with the results of my insomnia.

Adenovirus

Adenoviruses are a group of viruses which cause a variety of diseases in man and other animals, most of them fairly mild. They are being used for gene cloning applications in two ways.

- There is some interest in adenoviruses as gene cloning vectors for expressing large amounts of recombinant proteins in animal cells. Like many viruses, adenoviruses have the ability to 'turn on' their genes at a very high level. Adenoviral vectors seek to use this property by replacing a viral gene with another one, one that codes for a protein we want.

- The other interest is in using adenoviruses to make live virus vaccines. Here a protein from another, more dangerous pathogen is spliced into the DNA of a mild adenovirus. The foreign protein (which must not be dangerous in itself) is made whenever the virus infects a cell. Thus when the immune system makes the antibody to the virus, it also makes it to the foreign protein, and the individual becomes immune to that protein. A viral vaccine for rabies is in the early stages of development in the US.

See also **Viral vaccines**.

ADEPT (antibody-directed enzyme prodrug therapy)

This is a novel way to target a drug to a specific tissue. The targetting mechanism and the drug are administered separately. The drug is administered as an inactive prodrug that does not itself have any effect.

ADEPT (ANTIBODY-DIRECTED ENZYME PRODRUG THERAPY)

That prodrug can be converted into an active drug by an enzyme. Usually when a prodrug is used as a therapeutic the enzyme that converts it to an active form must be present in the body. However in ADEPT the converting enzyme can be, and indeed is preferably, one that does not occur normally in humans. Instead it is administered with a second injection. The enzyme is coupled to an antibody that concentrates it in the target tissue. When the enzyme has arrived at the target tissue, the prodrug is activated there to form the active drug, while elsewhere it remains inactive.

This is being developed for tumour treatment. The prodrugs being considered are prodrugs of highly toxic anti-tumour compounds, which in their normal form have severe side-effects because they kill many cells other than the tumour cells. Using ADEPT these drugs can be targeted to the tumour, sparing the rest of the body, by using an antibody that binds specifically to the tumour.

See also **Drug delivery**.

Affinity chromatography

This is a method for separating molecules by using their ability to bind specifically to other molecules. This method is of particular use in biological molecule separation because so many biomolecules bind very strongly and specifically to other molecules — their substrates, inhibitors, regulators, ligands, etc. (A ligand is the molecule, usually a small molecule or small molecular group, that binds to a larger molecule, usually a protein. The substrates of enzymes can be considered as ligands, as they bind to the enzyme, although they are not usually thought of in this way because, soon after they bind, they are converted to another molecule.)

There are two types of biological affinity chromatography. Either a biological molecule can be immobilized and a smaller molecule to which it is to bind can then be stuck to it, or the smaller ligand can be immobilized and the macromolecule stuck to it. (Of course, both sticker

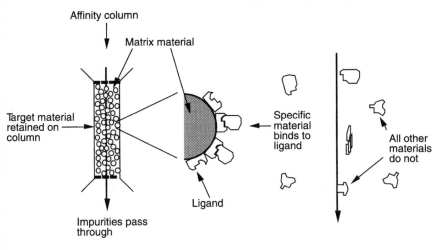

and stickee can be biological, too). A variant is to use an antibody as the immobilized molecule and use it to 'capture' its antigen: this is often called Immunoaffinity chromatography.

Biological molecules used to separate smaller molecules include:

- Enzymes, to isolate substrates (only works if at least one substrate is missing from the mix, otherwise the enzyme just destroys what you are trying to separate).

- Antibodies (for separating almost any molecule or molecular group from a complex mixture).

- Cyclodextrins (for separating lipophilic materials particularly).

- Lectins (proteins which bind specific sugars very strongly and specifically, and which are therefore used for separating carbohydrates and anything with carbohydrates attached).

A variation is pseudoaffinity chromatography, in which a compound which is like a biological ligand is immobilized on a solid material and enzymes or other proteins are bound to it. A range of complex organic dyes are very good at binding some types of enzymes (especially dehydrogenases) because of their similarity to the enzyme's 'real' substrate NAD or NADP — nicotinamide adenine dinucleotide or its phosphate. This is also called dye–ligand affinity chromatography. Other methods include metal affinity chromatography, where a metal ion is immobilized on a solid support: metal ions bind tightly and specifically to many biomolecules. The metal ion is bound to a chelator or chelating group, a chemical group that binds that metal, and usually only that metal, extremely tightly.

3

There are a wide range of support materials used in affinity chromato-graphy (*see* **Chromatography**).

To produce an affinity material, the solid support material to which the binding partner is going to be attached must be chemically activated. This is a process which takes the rather inert chemical material and adds a reactive chemical group to it, so that when the affinity-binding molecule is added to the support, it reacts with it to form a covalent link. Otherwise the affinity material would simply be washed off.

Affinity chromatography is widely used in research. It is also used in production, although the materials are usually too expensive for it to be useful for large-scale purification. It is used where a valuable product is to be isolated from a complex mixture of similar chemicals, of which the product is a minor component. Thus Armour Pharmaceuticals and Baxter Healthcare both isolate factor VIII (used to treat haemophilia A) from blood using affinity chromatography. An antibody is linked onto a 'column' of solid material, and the plasma passed over it: the factor VIII sticks and all the other proteins do not, resulting in a purification of roughly 200 000-fold.

Affinity tag

Also sometimes called a purification tag, this is a section of the amino acid sequence of a protein that has been engineered into the protein to make its purification easier. These can work in a number of ways.

If the protein is being produced as a fusion protein (i.e. several proteins made as one single polypeptide by the cell, needing to be chopped up afterwards by the biotechnologist), then a purification tag can be a short amino acid sequence between the 'units' of the fusion protein, allowing the protein to be chopped up easily. These could be the specific sequence recognized by a peptidase or protease, for example the Leu-Val-Pro-Arg-Gly-Ser sequence recognized by the enzyme thrombin (it cleaves between the Arg and the Gly).

The tag could be another protein, for example an enzyme (making the new protein easier to detect) or a protein which binds to some other material very tightly (such as avidin, which binds to the vitamin biotin very strongly), which would allow the protein to be purified by affinity chromatography. Enzymes sometimes fulfil both roles, as they catalyse the reaction of substrates and bind them to inhibitors very strongly. Short segments of cellulase (an enzyme which breaks down cellulose) have been used to make fusion proteins which stick to a cellulose affinity matrix.

The tag could be a short amino acid sequence, either random or selected from some other protein, which is recognized by an antibody. The antibody would then bind to the protein when it would not have done so before. One such short peptide, called FLAG, has been designed so that it is particularly easy to make antibodies against it.

The tag could be a few amino acids, which are then used as a chemical tag on the protein. A string of positively charged amino acids, for example, will bind very strongly to a negatively charged filter: this could be used as the basis of a separation system. Some amino acids bind metals very strongly, especially in pairs: this chemical property can be exploited by using a filter with metal atoms chemically linked onto it to pull a protein out of a mixture of proteins.

See also **Affinity chromatography**.

Agrobacterium tumefaciens

Agrobacterium tumefaciens causes a disease called Crown Gall disease in some plants. The bacterium infects a wound, and injects a short stretch of DNA into some of the cells around the wound. The DNA comes from a large plasmid — the Ti (tumour induction) plasmid. A short region of the Ti plasmid (called T-DNA, standing for transferred DNA) is transferred to the plant cell, where it causes the cell to grow into a tumour-like structure. The T-DNA contains genes which, among other things, allows the infected plant cells to make two unusual compounds, nopaline and octopine, which are characteristic of transformed cells. The cells form the gall, a hard lumpy growth on the plant, which is a benign home for the bacterium.

This DNA transfer mechanism has been harnessed as a way of genetically engineering plants. The Ti plasmid is modified so that a foreign gene is transferred into the plant cell along with or instead of the nopaline synthesis genes. When the bacterium is cultured with isolated plant cells or with wounded plant tissue, the 'new' gene is injected into the cells and ends up integrated into the chromosomes of the plant.

A. tumefaciens usually only infects some dicotyledons, because their response to wounding is compatible with *A. tumefaciens'* DNA transfer mechanism. When they are wounded, dicotyledons make specific phenolic chemicals that are part of their wound-protection mechanism. *A. tumefaciens* uses these compounds both as chemotactic agents (i.e. they swim towards the source of the compound, and so 'find' the wound) and to stimulate DNA transfer.

Monocotyledons do not respond in this way, and so are resistant to *A. tumefaciens*. This has been a problem in the past for biotechnologists as many agriculturally important plants, including grain crops, are monocotyledons. Manipulation of the plasmid and the conditions under which it transfers its DNA in culture have allowed grain crops (including rice and maize) to be transformed with T-DNA, but this is still a difficult technology to get to work.

A previous problem with *A. tumefaciens* was the size of the plasmid, which made it extremely difficult to handle using conventional recombinant DNA techniques. Now binary vector systems have been introduced to overcome this. The T-DNA is held on one small plasmid that is easy to manipulate *in vitro*. A much larger plasmid contains the *vir* genes which are necessary for the infection process, but which do not have to be manipulated. The two share enough DNA in common so that, when introduced into one cell, they recombine to form a single Ti plasmid containing the original *vir* genes and the newly manipulated T-DNA region.

A. tumefaciens has been used particularly to get DNA into trees. Trees are difficult plants to breed because of their size and long life cycles, and so genetic engineering techniques offer unusual advantages of speed and the ability to engineer millions of clones. Walnut, Poplar, Apple and Plum trees have all had DNA transferred to them using *A. tumefaciens*.

AIDS

AIDS (acquired immune deficiency syndrome) is the final stage of infection of man by human immunodeficiency virus (HIV). Infection is currently believed to be irreversible and almost always fatal, although how long it takes to die varies enormously between patients. Once thought to be caused solely by HIV, there is growing evidence that HIV alone cannot cause AIDS. In particular, it is believed that if someone has an infection of mycoplasma (a type of bacterium), then they are very much more susceptible to getting an HIV infection if exposed to the virus, and that cytomegalovirus (CMV), which many people carry around all the time, may trigger the change from an apparently harmless HIV infection to the full AIDS disease. There is also a theory — the HIVER theory — proposing that most of the damage caused by the disease is actually caused by an autoimmune problem, i.e. AIDS is the immune system destroying itself, triggered by the virus, rather than the virus being destructive. However the effectiveness of anti-HIV drugs shows that HIV must have an important role to play.

There are several areas in which biotechnologists have made progress in analysing the disease, in developing methods for diagnosis and treatment, and in working towards cures for and prevention of AIDS.

Fundamental research. The complete characterization of HIV was essentially complete within six years of recognizing that the disease existed, something of a record in medical history. That this could take place so fast is due to the techniques of molecular biology, and the ready availability of reagents to serve those techniques.

Diagnosis. AIDS is a very slow disease and HIV-positive people may be infectious but show no symptoms for years. Consequently there is a lot of interest in diagnosing HIV infection as soon as possible. A large number of tests based on monoclonal antibodies has been suggested, tried, and even developed, and many have been marketed. Others, based on DNA probes (*see* **DNA probes**), and especially PCR (*see* **PCR**), have been researched but are generally too complicated to be useful for large-scale clinical application.

Therapy. The only currently accepted AIDS therapy is treatment with AZT (Retrovir). This is a conventional chemical drug, which however could be made using biotransformation (*see* **Biotransformation**). A range of other drugs are in development, some based on the conventional pharmaceutical research done in the last few years. Others are biotechnological products, such as the CD4-based proteins, which aim to stop the virus from ever linking onto a cell, and so stop new infection. CD4 is the cell protein which the virus binds to. The protein gp120 (and its parent protein gp160) is the virus protein that does the binding. Covering either with some other protein will, in theory, stop the virus locking onto the cell. However CD4 is a membrane protein, so it is not soluble: consequently, one of the first goals of the recombinant DNA work is to make a soluble CD4. Genentech, Biogen, Chiron and many of the other 'big names' in biotechnology are pursuing variations on this type of AIDS therapy, although clinical trials have not been very promising to date for first generation soluble CD4's.

Vaccines. Developing a therapeutic vaccine for something that destroys the immune system is difficult. A prophylactic vaccine — one that protects people who do not have an HIV infection from catching the virus — should be easier to develop. Several approaches are being tried, based around the idea of cloning one particular protein, or part protein, from the AIDS virus and using it as a vaccine, thus avoiding having to inject HIV itself into people. The favoured proteins are the gp120 (or gp160) protein and proteins from the virus's core (p24), which, for some reason, seem to work quite well. No vaccine has come near large-scale clinical trials yet.

A substantial effect that the AIDS epidemic has had on the biotechnology industry is to speed up the regulatory process for some drugs. People with AIDS became fed up with how slow the official regulatory process was and started to try out drugs with potential effects on AIDS informally on themselves. A range of possible anti-viral compounds including interferon (which is not for sale as an anti-AIDS drug in the US) have been tried out by people with AIDS. This has in turn caused politicians to develop the 'fast track' drug approval processes for drugs for AIDS, and potentially for other terminal disease states.

AIDS has a high political profile (the 1992 'AIDS awareness' concert in memory of singer Freddy Mercury attracted a billion viewers, compared to around 250 million for the 'Live Aid' concert supporting African famine relief). Research (both academic and industrial) is intense. Funding for diagnostic and therapeutic products for AIDS has been easier to obtain than for many other diseases. The biotechnology industry has worked extensively on treatments for AIDS for three main reasons, the first of which is this comparatively easily obtainable funding. The second is the complex technical challenge of the disease, which attracts researchers to work on it. The third is the scale of the future problem: the 'Western' world probably has around 3 million people infected with HIV, most of whom will develop AIDS in the next ten years, and will want any effective treatment that the biotechnology community can produce.

Airlift fermenter

Airlift fermenters or airlift reactors (ALRs) are a type of loop fermenter, and are very popular in many applications. The fermenter consists of two parts, a riser and a downcomer. The liquid fermentation medium circulates between the two, driven up the riser by air (or other gas, sometimes pure oxygen) pumped into the bottom through a sparger. Thus there is no stirring mechanism in the fermenter. There is usually a gas separator at the top of the riser. This separates ('disengages') the gas from the liquid, so the bubbles of gas are not sucked back down the downcomer where they would try to rise, bringing the circulation of liquid to a halt.

The popularity of this type of fermenter is mainly due to the fluid dynamics of the reactor. The air drives the liquid around the fermenter quite gently, so reducing shear forces that might otherwise occur as stirring plates churn through the medium and break open delicate mammalian cultured cells, or damage long fungal hyphae. Airlift fermenters were very popular for making monoclonal antibodies in bulk, although the trend has now swung rather towards using hollow fibre reactors instead for all except very large-scale production.

See also **Hollow fibre, Loop bioreactor**.

Amino acids

The amino acids, key components of all living things, are produced by biotechnology in bulk using fermentation and biotransformation. Several Japanese companies dominate the world production of amino acids. They use fermentation systems in which they grow bacteria or fungi which have been selected to over-produce specific amino acids, which they excrete into the fermentation medium. Harvesting the medium and removing the other components yields amino acids in total amounts of hundreds or thousands of tonnes a year.

Commercially produced amino acids include:

- Glutamic acid. This amino acid is produced in larger amounts than any other, because it is widely used as Monosodium glutamate (MSG) in the food industry as a flavour enhancer, and in the Far East as a table condiment.

- Lysine. This is the second most abundantly produced amino acid, and is used as a supplement for animal feeds (which are often deficient in some essential amino acids, particularly lysine).

- Cysteine, methionine. These are the sulphur-containing amino acids, and again are used as animal feed supplements.

- Phenylalanine. As well as being used to a small amount as a feed supplement, phenylalanine is the most expensive chemical ingredient in the manufacture of Aspartame.

- Tryptophan hit the headlines in 1990 when the tryptophan produced by a new genetically engineered *Bacillus amyloliquefaciens* made by Showa Denko KK was linked to a rare degenerative disorder called eosinophila-myalgia syndrome (EMS). Despite loud and prolonged claims that this was proof that genetic engineering was dangerous, the problem was eventually traced to a chemical generated during the (perfectly conventional) purification procedure, and had nothing to do with recombinant DNA.

Several other essential amino acids, i.e. amino acids which our bodies cannot make themselves and hence have to be eaten in our diet, are also made in substantial amounts for human or, more usually, animal consumption. Indeed, most of the other 15 'natural' amino acids —

those found in proteins — are produced by fermentation in thousand tonne amounts. Other amino acids, not found in proteins, and especially the D-isomers, are made by biotransformations for use as chemical intermediates. Biotransformations are used for these materials because they are not found in nature, or are only found in tiny amounts. The D-amino acids, for example, are used in antibiotic manufacture. (D amino acids are those with the opposite 'handedness' to natural amino acids.)

See **Artificial sweeteners, Chirality**.

Animal cell immobilization

Animal cells are used widely in biotechnology to produce natural products or genetically engineered proteins. Animal cells have the advantage that they already produce many proteins of pharmacological interest, and that genetically engineered proteins are produced by them with the post-translational modifications normal to animals. However animal cells are much more fragile than bacterial ones, and so cannot be exposed to the high shear forces of repeated centrifugation which bacterial cells can tolerate in a commercial fermentation process.

Any cell (indeed, any small particle) can be immobilized by entrapping it in some solid material, either by having it grow there or by forming the material around it after it has grown. Entrapment in some form is common and widely used, from microencapsulation to growing cells inside a hollow fibre in a bioreactor (*see* **Hollow Fibre**). As well as these general approaches, there are some methods and materials which are specific to animal cells.

Surface adherant cells. The simplest is to use the natural adherance of animal cells to some materials. Many animal cells stick down flat on a suitable surface, hugging it as they would hug other cells or connective matrices in the body. If grown on suitable plastic surfaces, on glass or many ceramics these cells will stick to them. This makes them much easier to keep in one place. In general, between 10000 and 100000 mammalian cells will grow on $1\,cm^2$ of surface (the number depends rather on the surface and the type of cell).

This is a bulky way to grow cells unless the surfaces are folded in some way. Hollow fibre or membrane bioreactors can provide a way of doing that, but one of the favoured ways is to use porous carriers. These can be polysaccharide, protein (especially collagen), plastic or ceramic materials with microscopic holes in them a few tens or hundreds of microns across. Such materials are called microcarriers or microbeads. The cells grow into these holes, greatly increasing the surface area available to them while not increasing the bulk of the culture: the Opticore ceramic culture matrix, for example, has $8\,cm^2$ of surface per cm^3 of volume of solid material. The carriers can be formulated into small particles or into sheets or tubes. As well as ceramics, they can be made of polysaccharides (dextrans, alginates, agar) with various chemical modifications to give them a surface charge: these are popular because they mimic some aspects of the membranes on which cells grow in the body, and so the cells stick to them firmly.

Anti-idiotype antibodies

Anti-idiotype antibodies are antibodies which recognize the binding sites of other antibodies. Their binding sites are complementary to the binding site of another immunoglobulin. They are important to biotechnology in three ways.

Firstly, they occur in normal blood. When we become immune to something, we do not only acquire antibodies against that something. We also acquire antibodies against those antibodies (and antibodies against *those* antibodies etc.). This forms a network of antibodies which can all bind to each other to various degrees, a network which helps to regulate the immune response. It is possible that allergic responses are in part due to the breakdown of this sort of regulation. Thus anti-idiotype antibodies are important to the regulation of the immune system, and understanding how and why they are produced is an important part of understanding how the immune system works.

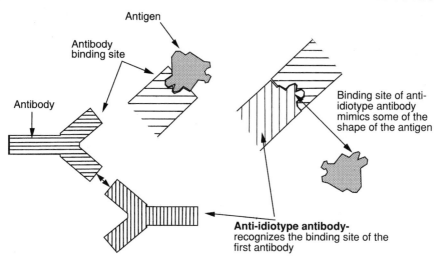

Another aspect comes from considering what an anti-idiotype antibody looks like. If an antibody is a 'key' exactly selected to fit the 'lock' of a virus or a bacterium, then an anti-idiotype antibody is a 'lock' exactly selected to fit that 'key'. In other words, it must have some similarity to the original antigen, the material to which the original antibody reacted. This means that making anti-idiotype antibodies could be a way of replicating the functional properties of such proteins as hormones or hormone receptor molecules. By raising an antibody against that molecule, and then raising an anti-idiotype antibody against the antibody, you will create an immunoglobulin with some of the functional characteristics of the original hormone or hormone receptor, but which can be produced more easily and which is chemically quite distinct.

Although this is fine in theory, antibodies only recognize a small region of the surface of a protein. Thus an anti-idiotype antibody can only mimic the properties or functions of that region of the protein, and these are likely to be rather limited. Thus, for example, an anti-idiotype antibody which binds to an antibody against insulin (and hence which will have a binding site looking like some of the insulin molecule) will sometimes bind to the insulin receptor molecule. However it does not necessarily make a cell respond in the same way as it would to insulin. This is because it may not bind to the receptor in exactly the same was as insulin itself binds. These relatively subtle differences have limited the use of anti-idiotype antibodies so far.

Anti-idiotype antibodies also have potential as vaccines. Here again, they are used to mimic a protein, in this case part of the surface of a virus or a bacterium. However it is not so critical in this case to mimic the whole viral coat protein. Providing that the anti-idiotype antibody mimics a part of the virus's surface which is readily accessible to the immune system (so that it is easily recognised in the final virus), then it could be use to stimulate the immune system to make a suitable antibody. This is a good idea because it allows a vaccine to be developed without ever using live virus to make it, so there is no risk of contaminating the vaccine with live virus. However the link between the virus being used to make antibody and the antibody, and between that antibody and the anti-idiotype antibody which is injected, and between *that* antibody and the antibody that our bodies are going to make, seems to be too tenuous. In the experiments done so far the resulting antibody fails to recognize the virus properly.

See also **Antibodies**

Antibiotics

Much of biotechnology is directed at discovering new drugs. One class is the antibiotics. There are three routes to developing new antibiotics (as opposed to improving the production of existing ones) with biotechnological components. Most new antibiotics are synthetic or semi-synthetic — it is very rare for a completely new type of antibiotic to be discovered in nature.

The 'established' antibiotics, especially penicillins, were the first products of the industry which is now biotechnology. Produced from fungi in fermentation systems, the penicillins, streptomycins and a host of other antibiotics were breakthrough drugs in the 1940s and 1950s, and are still major products of the fermentation industry. Since then, biotechnologists have built on this base to develop a range of new antibiotics.

Hybrid antibiotics. The synthesis of an antibiotic is the result of many enyzme steps in a bacterium or fungus. Some work is aiming at producing hybrid antibiotics — molecules which have bits from two different antibiotics. This is done by putting selected enzymes from two different antibiotic-producing cells into one bacterium. This work has gone furthest using genetically engineered Streptomycetes.

Novel metabolites. It is probable that there are vastly more antibiotics produced by microorganisms and plants than man has discovered yet. The biotechnology industry is exploiting its ability to grow new bacteria and fungi on a large scale to screen new bacteria for compounds which have useful drug activities. Xenova has specialized in this.

Animal antibacterials. Animals, especially invertebrates (which do not have the same sophisticated, adaptable immune system as mammals) produce a wide range of materials which kill bacteria. Most of these are proteins or peptides. 'Conventional' gene cloning technology can seek to clone the genes for such peptides into bacteria or yeast which can produce them in large amounts. Biotechnologists are particularly interested in proteins produced by cells of the immune system which normally destroy invading bacteria, and cells which produce proteins of the complement system, a set of proteins which knock holes in virus-infected cells. Some of these peptides do not destroy the cells themselves, but increase the chance that a white blood cell will destroy them (a process called opsonization). Others, such as the defensin peptides and the bacterial permeability increasing protein (BPI), bactenecin peptides, azurocidin, and the enzyme lysozyme actually kill bacterial cells. A third group typified by lactoferrin inhibits bacterial growth, in this case by removing the free iron which the bacteria need from their surroundings, and binding it in a tight, inaccessible complex.

Antibodies

Antibodies are proteins made by the immune system to combat infection. Each antibody is made to recognize one target antigen molecule. If the antigen is a small molecule, then the antibody will recognize all of it. If the antigen is a large molecule, then the antibody will recognize only a part of the antigen, which will be called that antibody's 'epitope'. The binding site of the antibody latches onto this antigen very specifically and very strongly. This latching-on allows the body to recognize the antigen as part of something that should not be there — a virus, a bacterium, a toxin — and so start the process of removing it.

Mammals make antibodies to almost anything that is a 'non-self' molecule, i.e. which one is not normally part of the body. Thus you can get a mammal to make an antibody against almost any molecule by injecting it into the bloodstream. The immune system recognizes it as a foreign material and makes a suitable antibody. In fact, it makes a whole range of slightly different antibodies: the blood of most people usually contains a vast host of different antibody molecules targeted at the various disease agents and other foreign molecules that have got into their blood in the past. For this reason the antibodies that are prepared from mammalian blood are called polyclonal antibodies, because they come from a large number of 'clones' (i.e. identical sets) of cells. This contrasts with the synthetic monoclonal antibodies (*see* **Monoclonal antibodies**).

Antibodies have been enormously useful to biotechnology because of their ability to latch tightly onto only one antigen, ignoring all others. For example, they would accurately distinguish sucrose from glucose, right-hand amino acids from left-hand (i.e. they distinguish enantiomers), human blood proteins from ape proteins etc. Thus they are the basis of many processes where great discrimination is needed.

Technically the antibody proteins are called Immunoglobulins. There are four types usually mentioned:

- IgM — the first type made by the body when it encounters a foreign material.

- IgG — the most common type, made after a prolonged encounter (as during a disease).

- IgE — the type responsible for allergic reactions.

- IgA — a rarer type which is present in saliva and some other non-blood fluids.

Antibodies are made by lymphocytes — they are actually made by B lymphocytes (B cells) in a process helped by T cells.

See also **Affinity chromatography, Antibody structure, Immuno-diagnostics, Immunotoxins**.

Antibody structure

Antibodies have a well-defined structure. Each antibody has two 'light' chains and two 'heavy' chains. The antigen binding region or binding site ('complementarity determining region') lies at the end of the light and heavy chains — it is therefore formed from both chains. The chains fold up into discrete blobs called domains: a 'single domain antibody' (Dab) is just one domain of an antibody.

The amino terminal domains of both heavy and light chains are called the variable domains, because they vary between antibodies. The other domains are constant domains, i.e. they are the same between antibodies of the same class and sub-class.

The antibody can be cut by proteases into several fragments known as Fab, Fab', and Fc (for historical reasons). These also feature in some biotechnological literature.

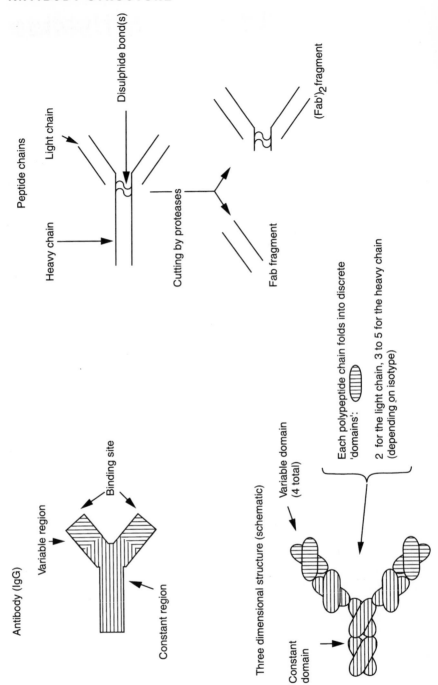

Peptide chains

Light chain

Disulphide bond(s)

Heavy chain

Cutting by proteases

(Fab')₂ fragment

Fab fragment

Antibody (IgG)

Variable region

Binding site

Constant region

Three dimensional structure (schematic)

Variable domain (4 total)

Constant domain

Each polypeptide chain folds into discrete 'domains':

2 for the light chain, 3 to 5 for the heavy chain (depending on isotype)

Antisense

Antisense RNA or DNA is a single-stranded nucleic acid which is complementary to the coding, or 'sense', strand of a gene, and hence is also complementary to the mRNA produced from that gene. If the anti-sense RNA is present in the cell at the same time as the mRNA, it hybridizes to it forming a double helix. This double helical RNA cannot then be translated by ribosomes to make protein. Thus antisense RNA can be used to block the expression of genes that make proteins.

Antisense RNA is a powerful way of modifying gene activity because it is a positive genetic engineering step, not a selection of negative mutants of a gene. Thus, rather than try to knock out all copies of a gene in, say, a plant, the genetic engineer only has to put in one gene which produces antisense RNA, and the antisense will prevent any mRNA from any copy of that gene being used by the cell.

How antisense actually works is still slightly mysterious. Clearly, ribosomes cannot use double helical RNA to make protein, so binding anti-sense RNA to an mRNA will block its action. However it rarely does so completely, suggesting that other factors are also involved. These may include:

• The way that the cell breaks down double-stranded RNA (many viral RNAs are double stranded, whereas normal cytoplasmic RNA is single stranded, so this may have evolved as an antiviral mechanism), and particularly the role of RNase H, an enzyme which specifically breaks down double-stranded RNA and RNA–DNA heteroduplexes.

• Where the cell antisense RNA is made (clearly it has to meet its target mRNA to be effective).

Antisense was discovered as the way that some bacteria naturally regulate the activity of some of their genes, but has been taken up enthusiastically by several companies to exploit its potential for artificially regulating genes. Antisense RNA, or derivatives of it, may be useful drugs, because they can block the effect of one gene without affecting any others. In particular, it could be used to block the effect of oncogenes (*see* **Oncogenes**), and so slow or prevent the development of cancer. It could also block the effect of viral genes, and so be used as an antiviral drug (*see* **Antiviral compounds**). Preliminary experiments show

ANTISENSE

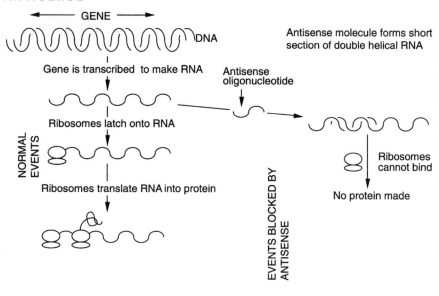

that antisense holds out great promise in these areas, and Genta and ISIS Pharmaceuticals are launching antisense pharmaceuticals into clinical trials. The major problem to taking that promise from experimental models involving cultured cells into more realistic animal models, is the problem of how to get the antisense into the affected cells. As the genetic engineering of humans is not practical, the pharmaceutical chemist has to be able to deliver intact antisense RNA or DNA to *all* affected cells. This is doubly difficult because RNA is quite unstable, and is very easily broken down by RNAses, enzymes found in very many tissues and very hard to destroy. One way round this is to use antisense DNA, or modified DNA (such as phosphorothioate DNA, which has one of the oxygen atoms in the phosphate groups replaced with a sulphur atom), which are more resistant to enzyme attack.

A more immediate application of antisense is in the genetic engineering of plants and animals. Plant genetic engineering in particular has benefited from antisense technology, where several groups have blocked the genes for specific enzymes. Most famously, the genes for polygalacturonidase have been blocked in tomatoes by several groups in industry and academe. Polygalacturonidase is one of the key enzymes used in breaking down the walls of cells in the flesh of ripening tomatoes, thus making them squashy. If a gene which makes antisense to the polygalacuronidase mRNA is spliced into the tomato plant, then its antisense product blocks the manufacture of that enzyme in the tomato, and the tomato stays firm much longer as it ripens.

See also **Ribozyme**.

Antiviral compounds

One of the areas in which biotechnology is playing a substantial role in developing new drugs is in producing antiviral compounds. The work has focused on a range of technical approaches.

One of the more established approaches is via a range of immune system boosters. Some, such as the interferons, are specifically antiviral. They stimulate cellular defences against viruses on many levels, from reducing cell DNA synthesis and so making cells more resistant to hijack by viral genes to enhancing cellular immune responses. The interferons, some of the first products of recombinant DNA technology, were originally hoped to be genuine broad-spectrum antivirals, but their value has been more limited to being used in combinations with other drugs as immune boosters, and in a few specialist applications

Biotechnologists have been very active in preparing complex chemicals with antiviral properties. The most popular line of approach is to make compounds which look like the nucleotides in DNA, which then block up the enzymes which allow the virus to make its own DNA without damaging the cell. Wellcome's AZT (Retrovir, the anti-AIDS drug) is one such nucleotide 'analogue'. These are complex compounds and must be synthesized in their correct stereoisomer if they are to work. Using enzymatic synthesis for at least part of their production can be useful. A range of enzymes which make part of the nucleotide molecules have been purified (phosphoryl transferases, glycosyl transferases, and enzymes which modify the bases) and are so efficient that they can work usefully fast on the nucleotide analogues even though these analogues are not their normal substrates. A range of nucleotide analogues, especially the 'carbocyclic' analogues (compounds in which the oxygen in the sugar ring is replaced by carbon), are being very actively investigated as potential antivirals for treating long-term viral diseases.

The second approach is to use genetic engineering to create proteins which block viral replication. The approach here depends on the virus concerned, but generally works by making a protein which binds to the protein on the cells which is the 'docking' protein for the virus, or to the virus's protein which is the 'docking probe'. In the former case a segment of the viral protein can do the trick, in the latter case a fragment of the cell receptor protein (*see* **AIDS**). Many other strategies have been suggested, but have not come to products near clinical trials.

The third approach is to use antisense RNA or ribozymes (*see* **Antisense, Ribozymes**). This is still very experimental.

See also **Biological response modifiers**.

Aquaculture

Aquaculture is the growing of water plants and animals in 'farms', rather than harvesting them from wherever they happen to grow in rivers or seas. A related and rather more specific term is pisciculture, the culture of fish. Usually aquaculture uses fresh water. When it does not, using sea water instead, it can be called mariculture. It is a peripheral part of biotechnology because it is a new commercial development, and hence uses the latest technologies rather than traditional ones, and because it often involves growing organisms in large volumes of water, which has similarities to growing large volumes of yeast or bacteria, biotechnology's 'heartland'.

Aquaculture is a growing industry, and produces a range of products.

- Fish, especially high-value fish such as salmon and rainbow trout. This requires fairly undemanding technology: fish-farming in some form was carried out by the Romans, and is the reason that why villages in England have a 'village pond'.

- Crayfish, lobsters, oysters, shrimps, and other molluscs. These are farmed even more intensively (i.e. with more animal mass per cubic metre of water) than fish, being even more stupid.

Biotechnology comes into animal aquaculture in providing clean, well-aerated water for the animals to grow in. It also often provides food, as krill, as powdered synthetic food, and as food additives such as astaxanthins (a bright red–pink pigment) to ensure that the fish and prawns have the right colour.

Aquaculture has also been used to mass-produce macro- and micro-algae (*see* **Biomass**). These are grown in the Far East not only for food but also for chemicals (agars and gums), vitamins, and pigments.

In both animal and plant areas, biotechnologists have been using genetic methods on aquacultured species, especially to produce triploid and tetraploid organisms, and hybrid algae through plant cell fusions. Triploid trout, for example, are sterile, and can be used for biocontrol of weeds without the threat of being able to breed themselves. Triploid oysters are reckoned by the US market to taste better than normal ones, and, being sterile, put more energy into muscle production and less into reproductive organs.

Artificial sweeteners

There are a wide range of substances used to make food taste sweeter without increasing its 'calorie value'. Among those of interest to biotechnology are:

Thaumatin. A protein produced by *Thaumatococcus danielli* in its fruit. Thaumatins are 3000 times as sweet as sugar, and at low concentrations enhance other flavours as well. Because they are proteins they can and have been produced by genetic engineering in bacteria, so avoiding having to go to the tropics to harvest the fruit. Thaumatins have been produced in *E. coli*, *B. subtilis*, *Streptomyces lividans* and *Saccharomyces cereviseae*, and the genes have been put into higher plants too.

Aspartame. Known as Nutrasweet, this is one of the most commonly used commercial artificial sweeteners. It is a dipeptide (aspartate–

phenylalanine–methyl). Because it is made from two amino acids, there are two parts of its manufacture which may be of interest to a biotechnologist. Firstly, one amino acid — phenylalanine — is relatively expensive, so selection, genetic engineering or other manipulation of a fermentation which produces phenylalanine more efficiently is a worthwhile goal as part of the production of aspartame. Secondly the synthesis of the dipeptide is achieved by enzymes: particularly by using a protease to join the two amino acids together (rather than its more normal reaction of separating them). Both areas are undergoing continuing commercial development.

Auxostat

An auxostat is a chemostat in which the dilution rate can vary. Usually a chemostat, a closed culture vessel into which new medium is continuously fed and old medium and organisms continuously removed, has a fixed rate of dilution, that is, a fixed rate at which the new material is added and the old removed. This determines how fast the organism in the chemostat will grow. In an auxostat the rate at which new material is added and old removed is determined by some feature of the culture. For example, the amount of bacteria may be measured by measuring the cloudiness (turbidity) of the culture, and the amount of material added adjusted to keep the turbidity constant. Alternatively, if the bacteria reduce the pH of the culture as they grow (as bacteria often do), the pH may be used to control the rate of dilution. The former is called a turbidostat, the latter a pHauxostat.

Auxostats have the advantage that the maximum growth rate or yield can be obtained much more easily than with a chemostat. If the dilution rate is not high enough in a chemostat, the culture will grow at less than its maximum growth rate. If it is too high, the organisms will not be able to keep up with the addition of new medium and so will be diluted out — you will end up with an empty chemostat. An auxostat can be adjusted to keep up automatically with bacterial growth, and so maximize the growth rate. At such a high growth rate bacteria which can grow fast are selected over ones which grow slowly. Thus natural selection acts on the population, selecting fast-growing variants of the bacteria. Depending on what the auxostat is to be used for, this can be a good or a bad thing.

In practice, all large continuous industrial fermentation systems are auxostats rather than chemostats, as they have many feedback controls which allow the operator to adjust the materials that the fermentation receives as it proceeds.

Bacteriophage

A bacteriophage is a virus which attacks bacteria. They have been used extensively in DNA cloning work, where they form the basis of convenient **vector** molecules. The bacteriophage (or 'phage') used most are derived from two 'wild' phages called m13 and lambda.

Lambda Phages are used to clone large pieces of DNA or RNA. They are 'lytic' phages, i.e. they replicate by breaking open (lysing) their host cell. If a few phages are spread on a host of bacterial cells, they lyse open the cells they hit, releasing more phages, which then lyse open the neighbouring cells, releasing more phage and so on. On a bacteriological plate, this results in a small clear zone — a plaque — where each original phage landed. In a bulk liquid culture it results in an enormous number of phage particles — 10^{14} per litre in some cases. Both plaques and bulk culture are useful sources of large amounts of the bacteriophage DNA, for analysis. Some lambda vectors have also been developed which are expression vectors.

The other principle bacteriophage vector is the m13 system. This phage can grow inside a bacterium as a plasmid, so that it does not destroy the cell it infects but causes it to make new phages continuously. It is a single stranded DNA phage, and is used for the Sanger di-deoxy DNA sequencing method (which requires single stranded DNA as a starting material). Messing has developed a well-known series of m13 vectors for cloning pieces of DNA into m13 for sequencing.

Both of these phages grow on *E. coli* as a host bacterium. Many other phages, both of *E. coli* and of other bacteria, are used in more specialist research applications.

Baculovirus

Baculovirus is a class of insect virus which has been used to make DNA cloning **vectors** for gene expression in eukaryotic cells. The vector system was derived from *Autographa californica* nuclear polyhedrosis virus to enable biotechnologists to make large quantities of proteins from cloned genes in insect cells (the cells usually used are a cell line derived from the fall armyworm). Baculoviruses have a gene which is expressed late on during their infection cycle at very high levels, filling the nucleus of the cell with many-sided bodies full of a protein which is not necessary to produce more viruses, but is necessary for the virus's spread in the wild. In the vector cloning system this gene is replaced by that which the biotechnologist wants expressed.

Production of the protein can be up to 50 per cent of the cells' protein content, and several proteins can be made at once so that multi-subunit enzymes could (in principle) be made by this system. However, were it not for one overriding feature, this system would still not be as useful as yeast or bacterial gene expression systems, because cultured cells from multi-cellular organisms (such as insects) are harder to grow than yeasts. The strength of the baculovirus system is that, as a genuine animal expression system, it produces proteins which are glycosylated like the proteins in animals. This, in combination with the relatively high levels of expression, can make this an attractive option for proteins which are to be used as biopharmaceuticals. In addition, baculovirus is non-infective and non-pathogenic to vertebrates.

The baculovirus DNA is very large (100–150 kb), and so conventional recombinant DNA methods cannot be used to engineer it. Instead plasmids containing the desired gene are recombined with the virus *in vivo* through homologous recombination

A novel use for baculovirus systems is as viral insecticides. A gene is inserted into the virus which is lethal to an insect (e.g. the endotoxin gene from *B. thuringiensis*), but does not affect isolated viral cells. This is then used to produce infective virus, which can (in principle) infect insects and kill them. There are technical problems with this, however (such as whether the virus is still infective in a real organism) as well as regulatory ones.

Binding

Much of biochemistry and molecular biology is about molecules 'binding' to each other. This means that they stick to each other because the exact shape and chemical nature of parts of their surfaces mean that they are 'complementary': a common model is 'lock-and-key', much used to describe how enzymes fit around their substrate. A crucial fact of biology is that many biological molecules bind extremely tightly and specifically to other molecules — enzymes to their substrates, antibodies to their antigens, DNA strands to their complementary strands and so on. This binding is entirely spontaneous, and depends on the chemical nature of the molecules concerned.

Binding can be characterized by a binding constant or association constant(K_a), or its inverse the dissociation constant (K_d). Mathematically, if two molecules (Mol-1 and Mol-2) link up to form a complex, then

$$K_a = \frac{[\text{Complex}]}{[\text{Mol-1}] \times [\text{Mol-2}]} \qquad K_d = \frac{[\text{Mol-1}] \times [\text{Mol-2}]}{[\text{Complex}]}$$

where [something] is the concentration of that something. For any given concentration of Mol-1 and Mol-2, the higher the K_a is or the lower the K_d is, then the more of the complex and the less of the free Mol-1 and Mol-2 there will be. In general, in biotechnology when someone talks about K_a or K_d, they want a very 'tight' binding, so the bigger the K_a or the smaller the K_d the better. Antibodies generally have K_a between 10^7 (bad) and 10^{10} (good). Hormones bind to the receptors with K_a values from 10^4 to 10^8. Proteins such as cytokines or growth factors can bind to *their* receptors much more tightly, with K_a 10^{10} to 10^{12}. The prize goes to streptavidin, the protein that binds biotin (*see* **Biotin**). The K_a for the Biotin–streptavidin binding is around 10^{16}, i.e. enough for streptavidin to be able to suck just 3 µg of biotin out of a small aircraft hanger-full of water.

Bioaccumulation

This is the accumulation of materials which are not critical components of an organism by that organism. Usually it refers to the accumulation of metal. Many organisms — plants, fungi, protists, bacteria — accumulate metals when grown in a solution of them. Sometimes this is part of their defence mechanism against the poisonous effect of those metals, sometimes it is a side-effect of the chemistry of their cell walls.

In a few cases, bioaccumulation is economically important as part of the microbial mining cycle. Using this biosorption process, metals present in very low concentrations in water can be accumulated in the cell walls of living organisms and subsequently harvested. Bioaccumulation and the use of bacteria to remove toxic metals from waste water as a purification (**bioremediation**) process are clearly closely related.

See also **Biosorption, Microbial mining**.

Bioassay

A bioassay is a way of measuring something that has as its key part some biological element. Usually it means a way of measuring the concentration of a chemical, although bioassays for magnetic fields (using homing pigeons or magnetic bacteria), ionizing radiation (measuring mutation) or other physical effects are possible.

Many bioassays have been in traditional use — the proverbial canary in the coal mine was a bioassay for poisonous gas, the canary being the biological element. Animals have been used extensively in drug research in bioassays for the pharmacological activity of drugs. However new bioassays are usually developed using bacteria or animal or plant cells, as these are usually much easier to handle than whole animals or plants, and cheaper to make and keep. Thus bacterial bioassays for BOD (biological oxygen demand) and poisons in general are in use in the water industry. Here bacteria are mixed in with a sample of water, and an instrument measures how well they can metabolize (and hence use up oxygen, produce carbon dioxide or, in one case, give out light). Many of the cytokines and growth factors that biotechnologists are now producing using recombinant DNA methods were originally identified using bioassays in which mammalian cells were used to detect miniscule amounts of the compounds concerned through their potent effects on the cell's behaviour.

On the border line between bioassays and chemical assays are immuno-assays and enzyme assays. These assays use proteins, made from a biological system, in an otherwise entirely standard chemical-type measurement.

Bioassays are no more convenient to use than any other chemical reaction, and so there is substantial work needed to turn bioassays into bio-sensors.

See also **Immobilized cell biosensors.**

Bioconversion

Bioconversion is the conversion of one chemical into another by living organisms, as opposed to their conversion by enzymes (which is biotrans-formation) or chemical processes. Synonyms are biological transformation or microbial transformation. Bioconversion has been used for a long time to make a few chemicals such as alcohol (from sugar) and more recently the drug ephidrine. However it is only since the Second World War that bioconversion methods have become commonplace.

The usefulness of bioconversion is much the same as that of **biotrans-formation** — especially its extreme specificity and ability to work in moderate conditions. However bioconversion has several different properties, among which are that bioconversions can involve several chemical steps. Bioconversion can also involve enzymes that are quite unstable, because the cell continuously remakes them as they break down.

The problem with bioconversion is that most bacteria either convert chemicals very inefficiently, in which case they are not much use to the biotechnologist, or they convert them very efficiently into more bacterium, which is also not much use. Thus to make an effective bioconversion process, the bacterial strain has to be optimized so that it converts substrate to product efficiently but does not convert the product into something else. This is a more difficult target than bioremediation or biomass conversion processes, and rather more difficult than microbial mining.

Bioconversions of a number of chemicals have been reported, and some are used commercially. A major commercial application is in the manufacture of steroids. The 'basic' steroid molecule, often isolated from plants, is itself a very complicated molecule, and not one that it is easy to modify by normal chemical means to produce the very specific molecules needed for drug use. However a variety of bioconversions that attack only specific bits of the molecule can be used. Bioconversion is particularly useful for introducing chemical changes at specific points in large, complex molecules such as steroids. In many cases bioconversion is used together with more traditional organic chemistry to complete a complex synthesis.

Other applications are **microbial mining** and **bioremediation**, the degradation of compounds which are difficult to attack chemically. A major class of such compounds are the hydrocarbons in oil, which bioconversion seeks to transform into more reactive alcohols and aldehydes. This can be done chemically, but requires extreme conditions and metalic catalysts, and usually results in a complex mixture of products. Bioconversion takes place at much milder conditions, and produces primarily a single product. Bacterial oxidation systems which can convert hydrocarbons to alcohols, aldehydes or acids are known in many bacteria such as *Pseudomonas oleovorans*. This soil bacterium has been the subject of a lot of work to make it more industrially efficient. *Pseudomonas* species contain a wide variety of plasmids which allow them to break down many organic chemicals, and so could be of use in bioconversion processes.

Bioconversion in organic solvents

Many chemical reactions which are to be targeted for bioconversion or biotransformation are conventionally carried out in organic solvents, not in water, either because the reagents are insoluble in water or because water causes undesirable side-reactions. Enzymes can also be used in organic solvents, but interest is growing in the use of bacteria in solvents other than water.

Some bacterial bioconversions can be carried out in mixed phases, where the bacterium is tough enough to survive living next to droplets of the solvent. This has the advantage that a large number of enzymes, or very unstable enzymes which would not survive on their own in a bioreactor, can be used for a bioconversion. The disadvantage is that the bacterium must be kept alive, and that bacteria produce all sorts of other metabolites than the one you are after.

See also **Organic phase catalysis**.

Biocosmetics

A term for cosmetics with a biotechnological ingredient, or an activity or action based on biological knowledge (rather than the cosmetic industry's empirical experience, or on marketing ploys). Since any cosmetic which has a substantial effect on skin physiology would be classed as a drug, and so have to undergo all the rigourous proof of efficacy and safety that drugs must undergo, biocosmetics must steer an exact course between being believed to be efficacious and not being too effective.

Biocosmetics divide into three areas: biomaterials, biologically based ingredients, and medically rationalized products. The last class include hypoallergenic products and UV-blocking agents, whose mode of action is supported by medical research but which are not 'biotechnological' as such. It also includes products such as liposome-based preparations which may or may not have the effects claimed for them, but whose use of biotechnology products or buzzwords undoubtably gives them a marketing edge .

Biomaterials include the use in cosmetics of collagen and collagen hydrolysate, a wide range of lipids as 'moisturizers' (including liposomes, which are claimed to carry active ingredients into the skin), fibronectin, and hyaluronic acid. These, and especially the last, are water-retention agents which are meant to prevent the skin drying and wrinkling. Lipids such as gamma-linolenic acid also have anti-inflamatory effects in some cases.

Biological ingredients include biotin, cyclodextrins, sphingosine and a range of pigments. These are all 'natural' products, i.e. are made in a living organism rather than by chemical synthesis, and so can be produced biotechnologically: however their actual effect is debated by medical professionals.

Biodegradable materials

Biotechnologists have preceded the 'green' bandwagon by some years in developing biodegradable materials. These endeavours generally fall into three areas.

- Developing organisms which can break down normal materials, especially plastics (*see* **Bioremediation**).

- Developing composite materials. Most 'biodegradable' plastics are composite materials of a conventional plastic laced with a biodegradable material such as starch. They break down when soil bacteria digest the starch, leaving small granules of the plastic behind. There is controversy over whether this is an improvement, especially as these materials are much weaker than unadulterated plastic, so you need to use much more of them to make bottles and containers of the required strength.

- Biopolymers. Most living things produce polymers to make cell walls or other structural materials. Some of these can be used to make things: however most are easily wetted, and tend to fall apart if left in the rain for a long time. A few exceptions have been found, the most well-developed of which are probably poly-hydroxybutyrate (PHB), developed by ICI and polycaprolactone (PCL). Both materials can be moulded like a normal plastic, and are water-resistant and watertight. However their structures can be attacked slowly by bacteria, and so after a period of months to years will be completely broken down. The only problem remaining is what to make out of it. (For demonstration purposes, ICI has made entirely biodegradable coffin handles — however this will not alter the waste budget of the Western world significantly.) Hundreds of tonnes of PHB are produced annually. It is adapted to a range of applications by mixing it with small amounts of the related polyhydrovaleric acid (PHV), another biodegradable polymer.

One strong, flexible, water-resistant, biodegradable polymeric material not usually talked about is wood. Quite a lot of plant biotechnology is aimed at trees, and biotechnologists are indeed working on the genetic engineering of trees (*see* **Agrobacterium tumefaciens**).

Biodiversity

This means the diversity of life in general, but has implications for the biotechnology industry.

Biodiversity is considered to be a Good Thing. If a country plants only one type of crop (for example), then a pathogen can sweep the entire country's crop from the fields. This happened in a wave of epidemics to the US wheat farmers in the 1960s. Thus planting more than just a single type (or cultivar) of a crop is protection against pandemics.

Biodiversity also applies to the larger world, where it examines the vast range of plants (and animals, although these are considered less important from a biotechnological point of view) currently alive. Many of them may produce something useful to man — a new drug, a new foodstuff, a new material. If they are allowed to be wiped out (and most of the species of plants live in the tropics, now under substantial threat), that potential will be lost for ever.

The role of biotechnology in this is double-edged. If biotechnologists produce a new wonder-wheat, then that will be planted in place of all other cultivars, and the wheat-growing world will end up being a monoculture — biodiversity will have been reduced. On the other hand, biotechnological methods are such that if you can transform one cereal with a gene you can transfer a lot more, so biotechnology could actually increase biodiversity by increasing the number of crops with desirable genes in them. And it has been argued that the 'green revolution', including biotechnology, has been so successful that farmers are no longer under pressure to grow monocultures of the most productive crop: indeed, many farmers in Europe are paid to let fields lie fallow to reduce production, and so would be under pressure to diversify.

On the rain forest question biotechnologists are less vociferous, but one of the key technologies in plant biotechnology is plant cloning, storage, and micropropagation, which could be used to store and propagate rare or endangered species.

Bioethics

Bioethics is the branch of ethics, philosophy and social commentary that deals with the life sciences and their potential impact on society. At one end it can be enormously useful in focusing attention on problems that need to be solved. At the other extreme it can become a name-calling argument between the 'pro-biotechnology' and 'anti-biotechnology' schools of thought which, as it reduces discussion to epithets and clichés, can make better sound bites. The American Human Genome Project has set aside around 3 per cent of its budget solely to consider ethical issues.The medical genetics and pharmaceutical communities have employed ethical experts for years. Thus the industry and regulators of biotechnology have a substantial concern in bioethics.

Bioethics is not strictly confined to classical ethics, but rather spills over into social policy and even politics. Issues of current concern are almost all to do with how far biotechnology should be encouraged or (as is almost always believed to be the case) prevented from doing something. These issues include:

- The validity of making animal models for human diseases (for example, **transgenic** models of cancer).

- The use and abuse of information about people's genetic make-up.

- The problem of trading off the testing of a new drug's potential side-effects with the need to have patients benefit from it as quickly as possible.

- The conditions under which recombinant organisms can be released into the world.

- The role of biotechnology in embryo and fetal research.

- The justification for patenting life forms.

Those who study bioethics have identified a number of common themes among the arguments that are brought to bear on biotechnological issues. The most controversial is probably the 'Yuk factor'. Others include the need for individuals' ability to determine their own fate, the need to protect the vulnerable from the unscrupulous, and so on, which are common to many wider ethical arguments.

There is also a strong component of public opinion in bioethics, although often the *reason* that people feel some particular way about the technology is not examined.

See also **Genetic information, Mythogenesis, Treatment Protocol Program, Yuk factor**.

Biofilms

A biofilm is a layer of microorganisms growing on a surface, in a bed of polymer material which they themselves have made. Biofilms tend to form wherever a surface on which bacteria *can* grow is exposed to some suitable medium and a supply of bacteria. Thus biofilms form on sites as diverse as domestic plumbing, hyperbolic cooling towers in power stations, sewage plants and teeth.

The bacteria stick to the surface by a combination of corrosion and glue. The Bacterial films are rarely a single type of organism — rather they are communities (or 'consortia') of different organisms. Some corrode the surface. This process, called biocorrosion, can go on on its own. It leaves the surface rougher and more chemically 'sticky'. Other bacteria synthesize extensive networks of sticky mucopolysaccharide polymers to stick themselves, and any other bacteria nearby, to the surface. The resulting films can be amazingly hard to get off. They also increase the surface's roughness (and so the amount of pumping needed in a pipe), and block the diffusion of oxygen through membranes.

The process of covering surfaces with biofilm in this way is called biofouling. It is a particular problem where liquid is recirculated around a closed loop of pipework (so any bacteria washed off the film can get the chance to stick back on next time round), or when filter membranes are exposed to bacteria. Unlike normal fouling of membranes by solids or large molecules, biofouling is an active process which, once under way, cannot easily be reversed by cross-filtering or reversing the flow through the membrane. The biocorrosion can also start to break down the membrane, making it leaky. Thus there is a lot of interest in the use of biocides (both in the solution and impregnated into the surface) to stop biofilm production.

Biofouling and biocorrosion can affect nearly all materials known. Bob Tatnell of the DuPont company estimates that about 50 per cent of all metal corrosion worldwide is caused at least in part by biocorrosion.

However, biofilms can also be used — some biosensors use a biofilm of cells to detect when the water flowing past them contains a poison, and biofilms growing on permeable membranes have been used to break down organic wastes.

Biofilms form rapidly where non-sterile water containing nutrients flows

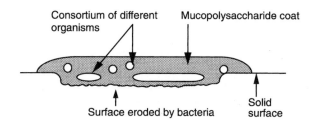

Consortium of different organisms   Mucopolysaccharide coat

Surface eroded by bacteria   Solid surface

over a surface — the slime forming on stones on the bottom of streams is an example, which also shows that, if the water is flowing fast enough, film cannot form. However, biofilm has been seen even when no obvious nutrient was available in ultra-pure water.

Biofuels

Biofuels are fuels made from bulk biological materials such as cane sugar or wood pulp. There is a range of ways of converting these rather bulky, inconvenient fuel materials into fuels which are useful for industrial or transport use, or as starting materials for the chemical industry. Converting biomass into replacements for gasoline has attracted most interest, especially since the 'fuel crisis' of the late 1970s.

Wet biomass materials, such as starch, sugar, sewage, waste waters etc. can be digested with enzymes or, more usually, by fermentation to make a variety of simple molecules, most usually ethanol or methane. Ethanol for use as a fuel is made from cane sugar by fermentation and distillation in commercial quantities in Brazil, where the economics are unusually favourable, and 'Proalcool' is a major fuel there: 14 billion litres were made in 1989. In the US various initiatives to promote 'Gasohol', various gasoline–ethanol mixes, have had mixed responses in the past, due to changing political support and a general discouragement from the oil industry. Most fuel alcohol made in the US is made by fermentation of corn (maize) starch. Methanol has also been suggested, but is harder to make and more corrosive. Methane is widely used as a heating fuel, and some biofuel methane has been tried out for electric power generation

The other gaseous biofuel is hydrogen, made primarily by photolysis of water. This is what photosynthesis does, but in normal living systems the hydrogen is not created as gas, but rather is used to make sugars. The aim of this area of biofuel research is to get organisms — probably single-celled algae — to release hydrogen gas when exposed to sunlight. This would be the ultimate clean, renewable fuel, but production has not approached the efficiency needed to make it commercial yet.

The other route to making biofuels is chemical. If any dry biological matter is heated up slowly, it undergoes 'pyrolysis' and produces a complex mixture of oily materials and charred polymers. These oils can be distilled in the same way that conventional mineral oil can be, to give fractions with similar properties to gasoline, diesel, lubricating oil, etc. The charred remains can themselves be burned, possibly to heat the pyrolysis reactors and stills. The chemical properties of the result can be rather different from conventional oils, and so far no one has succeeded in making this sort of process competitive with mineral oil production.

See also **Biogas, Solar energy**.

Biogas

Biogas is the name given to methane ('natural gas') produced by fermenting waste, and particularly sewage. This is an alternative method of waste disposal to landfill or conventional sewage plants. Waste is incubated with suitable bacteria in a digestor in the complete absence of air (an anaerobic fermenter). The organic matter in the waste is converted mainly into methane and carbon dioxide, and the methane can be burned to provide power, heat etc. In sewage treatment plants using anaerobic fermentation, methane is often used as a power source for the plant itself. The process is also called anaerobic digestion.

Anaerobic sewage disposal has some advantages over conventional systems (such as the activated sludge system). It produces less microbial mass to get rid of, and does not require aeration (which is expensive because it requires power). However it only works well on concentrated wastes: solid food waste or sewage sludge. Anaerobic fermentation is rarely a practical option for treatment of raw sewage, which is rather dilute.

The bacteria responsible for generating methane from waste are the methanogenic bacteria, an unusual group which can turn a limited number of carbon substrates into carbon dioxide and methane. To break the waste down into things that the methanogens can 'eat' requires other bacteria.

Thus an anaerobic digestor needs a specialized population of bacteria to work well. In practice waste digestion processes tend to use whatever organisms are on the waste, with consequently lower efficiency.

Biohydrometallurgy

This is the use of bacteria to perform processes involving metals. It encompasses a wide range of industrial processes, including microbial mining, oil recovery, desulphurization, and a range of physiological processes, including biosorption and the redox metabolism of bacteria. It is also the study of how bacteria corrode metal and metal-containing surfaces, a process known as biocorrosion.

In general, biohydrometallurgy contains two broad areas of bacterial activity.

Biosorption. This is the selective absorption of metal ions by bacteria and bacterial materials (such as their isolated cell walls).

Redox reactions. These are reactions where the bacterium uses the metal ion, or a mineral in which the metal is immobilized, for its metabolism. A major use is the oxidation of sulphide to sulphate, which reaction some bacteria use as a source of energy (the reaction releasing substantial chemical energy when carried out in air). As sulphides are frequently insoluble, and sulphates are often soluble, this is a fairly general method of releasing the metals in sulphide ores. The same reactions can be used to oxidize sulphides in one compound, releasing sulphuric acid which then dissolves another, or to pre-oxidize a metal ore to make it more amenable to further processing.

Bacteria can also oxidize or reduce metals themselves. The manganese nodules on the sea floor and the Banded Iron Formation rock strata (laid down 1000 million years ago) are probably the result of bacterial reduction of manganese and oxidation of iron respectively.

See also **Biofilms, Biosorption, Microbial mining**.

Bioinformatics

This is the use and organization of information of biological (almost always molecular biological) interest. In particular, it is concerned with organizing biomolecular databases, in getting useful information out of such databases, and in integrating information from disparate sources.

Among the most popular databases for molecular biologists are:

- DNA sequence databases. There are two main ones — GenBank (Los Alamos, USA) and the EMBL database (European Molecular Biology Laboratory, Germany). A Genome Project Database is being set up, partly in competition with these two.

- Protein sequence databases. There are two groups — PIR (Protein Identification Resource) in the US and MIPS in Europe, and the independent SwissProt database.

These two groups hold huge amounts of information about the sequence (of DNA bases and amino acids respectively) of natural genes and proteins. There are also databases about the three-dimensional structure of proteins (notably the Protein Database, run by Brookhaven National Laboratory in the US, which holds information on those proteins structures which have been determined by X-ray crystallography and, increasingly, NMR), and the structure of sugars, carbohydrates and glycoproteins. Databases about genetic maps (for the Genome Projects) and other genetic information are related to the DNA databases, and fall under the bioinformatics name. The US has set a National Center for Biotechnology Information (NCBI) at the National Institutes of Health to co-ordinate all these activities.

The major problem with these databases is not getting information into or out of them, but deciding what the information means. This is also an area in which information scientists are becoming increasingly interested.

Biolistics

This is a method, developed at Cornell University and commercialized by DuPont, to get DNA into cells. The DNA is mixed with small metal particles — usually tungsten — a fraction of a micrometre across. These are then fired into a cell at very high speed. They puncture it and carry the DNA into the cell. In the original system a 0.22 cartridge was used to drive the particles, hence this is also called a 'particle gun' system.

Biolistics has the advantage over transfection, transduction, etc. because it can apply to *any* cell, or indeed to parts of a cell. Thus biolistics has got DNA into animal, plant and fungal cells, and into mitochondria inside cells.

The force to drive the particles into the cells can also be electrical. A spark is used to vaporize a water droplet, which explodes like a small cartridge. This has the advantage that the current, and hence the energy of the explosion, can be varied at will. However, it is more complicated to set up.

As well as getting DNA into isolated cells, biolistics have been used to transfect DNA into animal tissues. Mouse skin and ears have been transfected by a suitably modified biolistic gun in whole, live mice, suggesting that this could be a route to somatic cell gene therapy in humans. The key to getting this to work is to limit the damage that the gun-like propulsion causes: curiously, tissue damage is not caused by the particles themselves, but by the blast of air or gas that accompanies them. The DNA was only active for a few days, however, before the cells broke it up.

See also **Transfection, transduction, transformation**.

Biological containment

This is restricting where genetically engineered organisms can go by arranging biochemical rather than physical barriers to prevent them from growing outside the laboratory.

Biological containment can take two forms: making the organism unable to survive in the outside environment, or making the outside environment inhospitable to the organism. The latter is rarely suitable for bacteria which, in principle, could survive almost anywhere. Thus for bacteria and yeasts the favoured approach is to mutate the genes in the bug so that they need to have a supply of a nutrient which is usually only found in the laboratory. If they get out, they then cannot grow. Other mutants may have weakened cell walls so they fall apart outside the laboratory, or may even have 'destructor genes' in them which destroy the cells if the temperature becomes lower or higher than the laboratory optimum.

Making the environment unfriendly to the organism is partly a biological control, partly a physical one. For example, some of the first genetically engineered rice strains were developed in England (which is too cold to grow rice) and tried in the field in Arizona (where it is too dry). Thus there was no rice growing nearby to cross-pollenate with the genetically engineered rice, and if any rice 'escaped' it would die. This is containment based on the biology of the plant, but without altering the plant specially.

Biological control

Also called biocontrol, this is the control of one species by another which has been deliberately introduced for that purpose. The most famous example is the introduction of myxomatosis into Australia to control rabbits, although biological control is much more ancient, dating back at least to the ancient Chinese who used Pharoah's ants to combat destructive insects in their grain stores.

BIOLOGICAL CONTROL

Biotechnologists have looked at many potential biological control agents: sometimes these overlap with biopesticides. For example, *B. thuringiensis* produces an anti-lepidopteric (caterpillar-killer) protein. *B. thuringiensis* has been used as a biocontrol agent for many years, but recently biotechnologists have isolated the protein responsible to make it into a biopesticide.

Biotechnologists approach biological control in several ways. Fungi, viruses, or bacteria which are known to attack a pest can be cultured in large amounts and sprayed onto a crop, there to kill that particular pest. Entamophagous fungi (fungi which infect insects) are favourites here, as they can infect insects through the cuticle, and so do not have to be eaten to be effective. Such fungi are termed mycoinsecticides, and about a dozen are under scale-up development.

Some mycoinsecticide fungi produce short epidemics (called epizootics) among the target pest population without creating a stable presence in the ecology: they can only continue to spread while there is a high density of the insect pathogen around and then die out.

In essence, culturing fungal pathogens is the same as culturing any other fungus, with the constraint that the fungi usually require very specific and unusual culture media.

Fungi, bacteria, and insects are also considered as weed-control agents: microorganisms to attack northern jointvetch and milkweed vine (weeds of rice and citrus trees respectively) are in use, and others under development.

Biocontrol can also be aimed at pathogenic fungi: Gary Strobel gained some notoriety in 1987 when he innoculated elm trees with a genetically engineered bacterium designed to protect them from Dutch elm disease without obtaining proper federal approval. Monsanto performed field trials of a bacterial biocontrol agent against the fungus that causes the wheat disease 'take-all' in 1988.

Biotechnologists have been more visible in producing viral biocontrol agents. Here genetic engineering and advances in culturing viruses in insect cells (*see* **Baculovirus**) have enable biotechnolgists to manipulate insect viruses to be, potentially, more effective biocontrol agents. The aim is to increase or alter the host range of a virus by altering the specificity of the viral proteins which bind to the cell surface, or to increase the virulence of a normally benign but very infectious virus by engineering in a toxin gene, or the 'pathogenesis' genes from another virus. In practice these aims are rather hard to achieve, as viral infection is a very complex process. Some trials have marked viruses with marker genes so that their spread can be monitored: this gives a measure of how well a simpler form of viral control — growing large amounts of the virus and then spraying it onto the crop — is working. Such field trials have been carried out, most famously in Scotland where pine trees were sprayed with an anti-moth viral biocontrol agent without (as it turned out subsequently) permission being given for release of this genetically modified organism.

The key to any biocontrol programme is to isolate an effective organism — one which can spread rapidly and effectively through the target pest population but which will not spread to other species (and hence become a pest in its own right). As pests are often foreign organisms introduced into an area where they have no natural enemies (e.g. water hyacinths in much of Africa, tumbleweed in the US, Dutch elm disease in many temporate climes), the best source of a potential biocontrol agent is often at the original home of the pest.

See also **Biopesticide**.

Biological response modifiers

A very general term, usually confined to mean proteins which affect how the immune system works. In this meaning it is almost synonymous with 'cytokine'. The term is widely used because of the existence of the FDA Biological Response Modifiers Advisory Committee, which oversees biopharmaceuticals which modify biological response mechanisms (i.e. all of them to date). Normally biological response modifiers act in concert, not as isolated chemical entities. Thus there has been much agonizing about how the components of biological response modifier drugs, cloned as pure proteins but used in combinations, can be regulated by the drug regulatory agencies, and particularly by the FDA. Cetus had well-publicized problems trying to get its interleukin 2 (IL-2) approved as an anticancer drug, because IL-2 was not effective on its own. Cetus wanted to use it in combination with other biopharmaceuticals, but as it was not effective enough as an isolated entitity it was refused approval. (Cetus admitted afterwards that their case was not helped by their inexperience at presenting data to the FDA.)

Biomass

Biomass means the mass of organic material in any bulk biological material, and by extension any large mass of biological matter. Single cell protein (SCP) technology is a version of biomass, but usually this term means growing plants (any plant from single-celled algae to sugar cane) and cropping it without very sophisticated processing to make a plant-derived food for people, animals or chemical processing.

Biomass has split into several areas of interest.

SCP. Single Cell Protein (*see* **SCP**).

Algal biomass. Single celled plants such as chlorella and spirulina are grown commercially in ponds to make food materials. Spirulina enjoyed a vogue as a health food a few years ago, due to an unfounded belief that it was incredibly nutritious. Like most algae (including some seaweeds) it is quite a good food, but spirulina is not outstanding. Chorella is grown commercially to make into fish food: it is fed to zooplankton (microscopic animals), and these in turn are harvested to feed the fish in fish farms. This is a way of converting sunlight into food in a more convenient and controllable way than normal farming.

Plant biomass. Crop plants such as sugar cane have also been grown for biomass. This is usually used as the start of a chemical production process (as growing plants for food is usually called 'farming'). Brazil spent a considerable amount of effort and money growing sugar to make ethanol by fermentation — relatively unprocessed sugar cane was used as the substrate, and the product used to run cars. This is the use of biomass as a way of converting sunlight into useful chemicals.

See also **Biofuels**.

Biomaterials

'Biomaterial' is a general term for any biologically-derived material which is used for the sake of its material property, rather than because it is a catalyst or a pharmaceutical. Thus DNA could be a biomaterial if you used it to make paperclips or cranes, rather than using it to store information.

The most common biomaterials are some proteins, many carbohydrates, and some specialized polymers. The proteins considered for biomaterials applications are usually proteins used as structural elements in animals or, occasionally, plants. Collagen, the protein in bone and connective tissues in a wide variety of animals, is a common candidate, and has been used (controversially) as a cosmetic biomaterial, being used as a 'natural filler' for plastic surgery operations. Fibroin, the protein in silk, has been suggested as a protein with sufficient strength to rival nylon or even Kevlar as a structural material. Most of these structural materials have fairly simple amino acid sequences, as they are made of short blocks of amino acids repeated many times. Thus the rigid, central sections of the collagen molecule, which give it its elastic strength, are made mostly of repeats of the three amino-acid unit glycine-X-proline (where X can be one of several amino acids). Biotechnologists have therefore made synthetic proteins with simple repeating patterns in the search for new biomaterials.

Carbohydrates have been used as structural materials for millenia: the strength of paper and papyrus derive from the properties of their carbohydrate, principally their cellulose, components. Biotechnology has produced a wide range of carbohydrates with modified properties which act as lubricants for biomedical applications, or as texture-modifying or bulking agents in food manufacture. These include rare but natural materials made from bacteria such as poly-dextrose, carbohydrates modified by enzymes to have improved properties, and entirely artificial polymers.

Other polymers include 'natural' plastics such as polyhydroxybutyrate (*see* **Biodegradatiable materials**), or rubbers produced by bacteria or fungi.

The properties of a polymer which are crucial in determining whether it will make a 'good' biomaterial for a specific application include:

- Tensile strength (both elasticity and breaking strength).
- Hydration (how much water does it bind? How much does it need to bind to keep its properties?).

BIOMATERIALS

- Viscoelastic properties.

- Viscosity.

See also **Biomineralization, Wood.**

Biomimetic

Literally meaning 'imitating life', this means the areas of chemistry that seek to develop reagents that perform some of the functions of biological molecules. The reason for doing this is that many biological molecules are chemically inconvenient to produce, handle or apply in large amounts and using cheap processes. By using chemical mimics of them the biotechnologist hopes to achieve a more flexible and more commercially useful way of achieving the same ends.

Areas of chemical research in the general field of biomimetics include the following.

Cofactor substitutes. Many enzyme co-factors are complex and labile molecules: in particular NAD and NADP (nicotinamide adenine dinucleotide, and NAD phosphate respectively) are difficult to work with on a large scale. Two lines of research seek to replace them with other molecules. Triazine dyes have been used as replacements for NAD in affinity chromatography applications. Here the dye is bound to a column, and a mix of materials containing a dehydrogenase enzyme is passed over the column. The triazine dye binds to the dehydrogenase as NAD would, and so holds it onto the column — all other materials pass through. This has been used very successfully for many purifications. The other application of cofactor substitutes is as actual substitutes for the substrates, especially for NAD, NADP, and FAD (flavine adenine dinucleotide) in reactions catalysed by dehydrogenases. Here again the aim is to find a small molecule that will do the chemical work of NAD etc. with the enzyme.

Peptide and DNA substitutes. Peptides and DNAs are rapidly broken down in many biological situations. Biotechnological chemists are working on altering the basic 'backbone' of peptides and nucleic acids so as to make

48

them more stable and potentially easier to make. In early 1992, for example, a 'DNA' substitute was reported that had no sugar–phosphate backbone at all: in its place was a polyamide chain looking more like a protein. This material bound tightly to single-stranded DNA in a way that suggested that it was forming correct base pairs. This has applications in **antisense**, as such molecules would be much easier to get into cells and totally resistant to breakdown by nucleases or proteases.

Synzymes. These are low molecular weight molecules that act as artificial enzymes, i.e. as highly specific catalysts. Usually they are synthesized to deliberately copy the three-dimensional structure of the 'active site' of an enzyme, but using non-peptide chemical building blocks. Unlike more common organic chemistry catalysts, which catalyse a broad range of reactions, the aim is to make synzymes as specific as enzymes.

Molecular imprinting. This is another approach at the same idea of getting a non-biological chemical to imitate some of the properties of a biochemical. In this case, a polymer material is 'imprinted' with gaps that exactly fit one and only one species of small molecule, in the same way that the binding site of an antibody exactly fits its antigen. This is done by forming the polymer matrix in the presence of the small molecules, so that the chains fold around those molecules. The small molecule is then washed out using suitable solvents, leaving 'holes' behind in the polymer. These can have quite a high affinity for the molecule that has been washed out, and so could be used to purify those molecules away from many others. In addition, just as antibodies that are raised against a transition state analogue can have catalytic activity (i.e. can be catalytic antibodies), so an imprinted polymer that has gaps that are 'shaped' to fit a transition state analogue can be a catalyst.

Biomineralization

This is the deposition of minerals by living organisms. In some applications it is related to microbial mining (the breakup of minerals by microorganisms) and hence is part of biohydrometallurgy. However biomineralization extends beyond this. There are two general areas of interest to the biotechnologist.

- Microbial biomineralization. This is the laying down of minerals by microorganisms. If the minerals are deposited inside the bacterial cell, then they are of necessity laid down as extremely small crystals or granules. The magnetite made by magnetic bacteria is of this sort — this magnetic mineral is made as tiny inclusion bodies inside some bacteria, which are thereby enabled to swim preferentially along magnetic field lines. (This enables them to swim towards the bottom of ponds in temperate zones.) Many larger mineral forms are also made partly by bacteria, and this has been suggested as a way that minerals could be isolated or purified using biotechnology.

- Multicellular biomineralization. In many plants and animals minerals are used to give strength. Thus vertebrate bone often contains calcium phosphate, and many grasses have silica in their leaves to give them a hard cutting edge to dissuade animals from eating them. The control of biomineralization is of substantial interest in several human diseases, notably osteoporosis, a disease in which too much calcium and phosphorus is lost from the bones.

Biomineralization is of interest to the materials scientist as well. Biological systems manage to deposit minerals in unusually useful forms. Thus bone and teeth are much stronger than 'raw' calcium phosphate. Additional strength and specific crystal forms are potentially useful as ways of extending the range of mineral materials available for construction, electronics, and chemical industries. Living things achieve these feats by incorporating specific proteins into the growing mineral, to force crystal growth into the form they require, or to reduce the propagation of cracks through the mineral when it is stressed.

Biopesticide

A biopesticide is a pesticide, i.e. a compound that kills animal pests, which is based on specific biological effects, and not on broader chemical poisons.

Specific types are also called bioinsecticides and biofungicides. Biopesticides are different from biocontrol agents in that biopesticides are passive agents, similar in concept to any pest control chemical such as a herbicide, whereas biocontrol agents are active, living things that seek out the pest to be destroyed.

There are a wide range of materials which plants produce to foil pest attack — the caffeine in coffee beans is probably such a material. However the most attractive for biotechnolgists are protein antipest materials, such as the much-hyped *Bacillus thuringiensis* (*B.t.*) toxin (sometimes called B.t.k., because it is *Bacillus thuringiensis* toxin type *k*), which specifically interferes with the absorption of food from the guts of some insects but is harmless to mammals. This protein (which has been used as a pesticide for some time as a bacterial suspension) has been 'cloned' into more amenable bacteria. The gene for the protein has also been inserted into petunia by Calgene, making a plant that is more resistant to pest attack.

The rationale behind developing biopesticides, by contrast with conventional pesticides, is twofold. Firstly they are more likely to be biodegradable than chemical entities which are not normally found in nature. Secondly, they are aimed to be more specific (and sometimes, as a consequence, more potent), as they are targeted at specific elements of the pest's metabolism.

Biological control agents are sometimes also known as biopesticides. By the end of 1991 there were 45 biopesticides or biocontrol agents aimed at insects (mostly bacteria, bacterially-derived proteins, or viruses), 10 at organisms which cause plant diseases, and 2 at weeds.

See also **Bacillus thuringiensis, Biological control**.

Bioreactor

A bioreactor is a vessel in which a biological reaction or change takes place, usually a fermentation or biotransformation. Bioreactors, and indeed fermentation and biotransformation, are central to much of biotechnology — everything from baking bread to producing genetically engineered interferon takes place in a fermentation, and hence uses a bioreactor.

51

BIOREACTOR

Bioreactors are conventionally divided up into three size classes. Laboratory bioreactors cover bench-top fermenters (up to 3 litres volume) and larger, stand-alone units (up to about 50 litres). These are used for research, and are usually used to create the fermentation process. Pilot plant fermenters are used to scale up a fermentation process, and to optimize it. They are typically between 50 and 1000 litres. Pilot plants have to be quite flexible to allow for process optimization. Production units can have any capacity, but usually hold at least 1000 litres, and can go up to the 1 000 000 litres of the ICI Pruteen plant. They generally are much more specialized than pilot plants, being designed to operate one process with maximum efficiency.

One of the key limiting factors in fermenter operation for more than a few litres of culture medium is oxygen availability: how to get oxygen to the growing organisms fast enough. Oxygen is poorly soluble in water, and so the liquid itself holds very little, which the organisms in a dense culture can use up in a few seconds. Thus they must be constantly supplied with oxygen gas, either as pure oxygen (efficient but expensive) or as air. Generally gas is bubbled through the reactor liquid: the smaller the bubbles the more efficiently the transfer of oxygen from bubble to solution (and hence to microorganism). However creating small bubbles requires power, may cause disruption of the organism growing in the liquid, and can cause foaming which fills the reactor vessel up with viscous foam. Anti-foaming agents can help this latter problem (which is also a problem when the

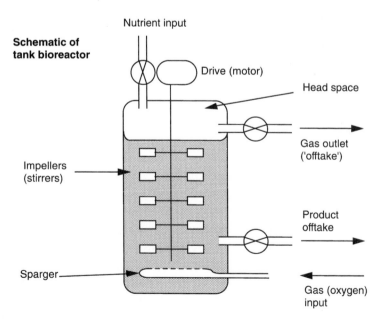

Schematic of tank bioreactor

Nutrient input

Drive (motor)

Head space

Gas outlet ('offtake')

Impellers (stirrers)

Product offtake

Sparger

Gas (oxygen) input

organisms produce a lot of carbon dioxide gas). The stirrers, spargers, loops, etc. referred to in other entries on fermentation are there primarily to increase oxygen uptake by the reactor liquid.

There are a number of separate entries about bioreactors (*see* **Hollowfibre, Immobilized cell bioreactor, Tank bioreactor**). They cover different types of bioreactors:

- Tank bioreactors (which is most of them).

- Immobilized cell bioreactors.

- Fibre and membrane bioreactors.

Other simpler types of reactor are not covered specifically. These include pond reactors and tower fermenters. Ponds are simply — ponds: they are used mainly for growing algae. Tower reactors are relatively simple towers in which nutrient is injected at the base and product collected at the top. They may be operated in batch or in continuous mode. They are used typically for anaerobic fermentation, i.e. fermentation where no air is needed, as for example in brewing.

A general type of reactor is the plug flow reactor. Here a substrate flows past a plug of solid support material, emerging from the end changed by the plug. Often the whole is in a pipe. The solid support can contain an enyzme or an organism. This is in fact a bioreactor equivalent of column chromatography.

See also **Biosensors, Chromatography, Fermentation processes, Fermentation, Substrates, Scale-up**.

Bioremediation

Bioremediation is the use of biological systems — almost invariably microorganisms – to clean up a contaminated site ('environment'). Sewage treatment plants perform this in a limited way: bioremediation covers use of microorganisms to destroy more toxic materials than are usually found in sewage, and to destroy them *in situ*, usually in soil or in waste tips.

BIOREMEDIATION

The basic twofold approach to most bioremediation projects is:

1. Selection of the microorganism. Soil which has been contaminated with the target chemical for some time is a favourite site to find an organism capable of breaking down that contaminant. Often such soil is found near a pipe junction or tank overflow valves in the plant which makes the chemical. Variants of this organism which grow faster or degrade the target chemical more efficiently are then created in the laboratory, by combinations of traditional microbial genetics, recombinant DNA methods or selection. Typically bioremediation methods use a consortium of organisms, rather than a single organism, which can catalyse the breakdown of different components of a pollutant or can perform different parts of the breakdown of a complex molecule. Even so, some molecules are quite refractory — PCBs can be dechlorinated by obligate anaerobes (bacteria killed by oxygen), and the carbon skeleton broken down by aerobes (organisms needing oxygen): however, clearly these two cannot work together at the same site.

2. Inoculation of the environment. The microorganism is introduced into the site, usually with a nutrient mix to support its growth and encourage it to break down the target compound. Oxygen is usually a limiting factor, as most targets for bioremediation are complex hydrocarbon-based compounds which must be metabolized by oxidation: nitrogen and phosphorus are also commonly added so that the bacterium's growth is limited by the availability of carbon. Thus the bacterium is under continued selective pressure to use all the carbon available in the soil for growth, including that present in the target compound. This phase of bioremediation, as critical as identifying a suitable microorganism, requires a substantial knowledge of microbial physiology and ecology. The main cause of failure of practical bioremediation projects is that the organism selected cannot perform the breakdown at a useful rate at the site, despite performing well in the laboratory. Clays, for example, are particularly poorly suited to bioremediation: as they are very densely packed, water penetrates them very slowly and air hardly at all.

Typical target compounds are chlorinated aromatics (although disposing of PCBs has met with only limited success), vinyl chloride, solvent residues, gasoline fractions, and crude oil. Alpha Environmental has hit the headlines on several occasions with its oil-eating bacterial preparation, used to digest oil spills at sea into soluble molecules which other bacteria can digest. Its most public application was in the Persian Gulf in 1991. This, the breakdown of compounds into biomass, is a type of biodegradation. Other, non-organic materials can also be metabolized if their end-product is nontoxic or volatile: selenium has been removed from soil by conversion to volatile compounds or elemental selenium, and nitrates have been removed from sewage waste by biological reduction to nitrogen gas for decades.

If a target site is very highly contaminated, or too cold or dry for bacteria to flourish in, then the soil can be placed in a tank bioreactor and the bioremediation carried out there. These bioreactors are essentially large insulated tanks into which soil or waste is placed with a bacterial inoculum and air blown to keep the mass oxygenated. Peter Wilderer at Hamburg has used a biofilm-based tank reactor to remove aromatic hydrocarbons — specifically benzene, toluene and xylene, the BTX mixture — from landfill site leachate. A film of organisms growing on a permeable membrane was used in order to capture the volatile hydrocarbons from the water.

Biosensors

Biosensors are devices which use a biological element as an intimate part of a sensor. For example, an electrode could have an enzyme immobilized on its surface so that it generated a current or voltage whenever it encountered that enzyme's substrate. There are several classes of biosensor:

- ISFET (ion-sensitive field-effect transistor) based devices.
- Physical sensors (including sensors for mass and for heat output).
- Enzyme electrodes.
- Immobilized cell biosensors.
- Immunosensors (*see* **Immunosensors**).
- Optical biosensors.

Other biosensors use DNA probes as the biological element, or even multicellular organisms such as daphnia (a small fresh water shrimp) or trout.

Biosensors have the potential for being extremely sensitive and specific ways of detecting something. However their practical application has been hampered by the biological element being very prone to destruction by whatever is meant to be detected. Thus, while for commercial applications, a sensor system should be either very cheap and disposable or able to operate continuously for some time, nearly all biosensors are very difficult to make in bulk and last for only a few measurements. The major problems found were:

- Stability. The biological element 'went off' quite rapidly with use. Some went off in minutes when operational requirements were for days or weeks of operation. Papers on biosensors often claim stability for weeks of operation, but this usually means that they are used once a day and kept in the fridge in between uses, a far cry from being used in a production line 24 hours a day.

- Shelf life. Even when they were not operating, the electrode goes off unless stored in a fridge or (in extreme cases) a freezer. This is useless if it is to be sold in a conventional shop.

- Manufacturability. Most biosensors are very difficult to make, and constructing an assembly line to make them in commercial quantities requires a well defined way of making them. Even commercially successful sensors are hard to make sufficiently reliably for such a method to be defined.

The most prominent exception is the glucose biosensor, an enzyme electrode based on glucose oxidase and commercialized by several companies, notably Exactech, as a test for blood glucose levels. These work where others fail because the amounts of glucose being measured are large (so the electrode does not have to be very sensitive) and the enzyme glucose oxidase is unusually stable.

Biosorption

Biosorption is the sequestering (i.e. capture from solution) of chemicals (usually metals) by materials of biological origin. Biosorption is widely talked about and little used as a method for removing materials from waste or for purifying rare metals.

Many organisms have components which bind metal ions: human bone matrix material, for example, binds strontium rather well. In some cases this is an active process — the organism uses energy to take the metal ions inside and trap them in an insoluble form. In others the process is passive — the metals stick of their own accord to a material that the organism makes. In both cases organisms can be selected which can accumulate more of the 'target' metal, or which accumulate one metal specifically. For industrial use, bacteria or yeast are almost always the organisms used, although many other organisms such as protozoa, simple plants, even trees can accumulate substantial amounts of metals.

Among the ways in which organisms actively accumulate metal ions is the precipitation of them as phosphates or sulphides by 'pumping' them into special sections of the cell. 'Passive' systems include proteins which bind the metal specifically (metallothioneins, for example — sulphur-containing proteins found in many organisms), lignin (from wood), chitin, chitosan, and some cellulose derivatives.

Biosorption is a biological phenomenon, and is interesting for its insight into how organisms cope with metal poisoning, lack of essential nutrient, etc. It can also be adapted to direct industrial use as a purification system by immobilizing the organisms on filters or in pellets, by using a recycling reactor system which passes the water to be treated through a bed of bacteria in a fermenter, or by extracting the biosorptive material from the organism and using that on its own. This latter option allows non-microbial biosorption systems: chitin, for example, absorbs a number of metal ions and is produced from waste prawn shells.

One of the most common waste-removal targets is removing heavy metals from industrial waste water, especially nuclear waste streams, where the metals are present at low concentrations but are the most hazardous element in the water. There is also substantial interest in using biosorption to purify precious metals such as silver and gold from very low grade ores, by washing the metal out of the ore and then concentrating it from the leachate using biosorption.

To be useful, biosorption must be specific and efficient. For metal removal from waste streams, removal must be at least 90 per cent efficient to be of any industrial use, and the organisms or polymers must be able to remove at least 15 per cent of their own weight in metal. Any less efficient system costs more to use than conventional removal systems (such as 'ion-exchange' materials). The efficiencies for precious metal extraction can be lower, depending on how valuable the metal is, but must be very specific: there is no point purifying gold if you purify a lot of lead along with it. As well as being improved by breeding and selection systems, biosorption can be improved (in principle) by genetic manipulation, by altering the structure of metal-binding proteins such as the metallothioneins, or by altering the enzymes which make other materials such as the chitosans or lignins. However, although it is talked about a lot, biosorption is not usually understood well enough to make such genetic engineering feasible yet.

Biotin

Biotin, a natural coenzyme, turns up in some unexpected places in biotechnology as a 'label' system. Biotin can be linked onto many different macromolecules by chemical reaction, a process called biotinylation. The protein avidin (usually made from egg white) or its bacterial counterpart streptavidin binds to biotin extremely tightly — far tighter than an antibody binds to its antigen. The avidin can be labelled with an enzyme, a fluorescent group, a coloured bead, etc. This will then seek out and recognize the biotinylated molecules, and not stick to any others. This can be preferrable to trying to link the enzyme, fluorescent tag, or other label onto the target macromolecule directly because (i) you can get more biotins onto a macromolecule than enzyme molecules, and (ii) the biotin is very stable, and so can be treated with extreme pH, boiled, or irradiated whereas an enzyme would be destroyed by these conditions.

Biotransformation

Biotransformation is the conversion of one chemical or material into another using a biological catalyst: a near synonym is biocatalysis, and hence the catalyst used is called a biocatalyst. Usually the catalyst is an enzyme, or a whole, dead microorganism that contains an enzyme or several enzymes. The advent of catalytic antibodies and ribozymes will broaden the definition somewhat. Conversion of one material into another using whole living organisms is usually called bioconversion.

Biotransformation is one of the largest areas of applied biotechnology (as opposed to research technologies): around 5 per cent by volume of the enzymes used industrially are used for biotransformation (nearly all the rest are used in the food industry, or in detergents). A wide range of materials are made by biotransformation, from commodity items such as high-fructose corn syrup to speciality chemicals for the pharmaceutical industry. Some biotransformation processes, such as that producing vitamin C, produce thousands of tonnes of product per year. The advantage of biotransformation over conventional chemistry is the specificity of enzymes. Reactions can be:

- Stereospecific — i.e. they produce only one optical isomer of a chiral compound.

- Regiospecific — i.e. they change only one part of a large and rather homogeneous molecule (analagous to only digging up one particular stretch of a motorway).

A key use for biotransformation is in 'resolution'. This is a biotransformation which takes a racemic mix of a chiral compound and converts one optical isomer into another compound. This means that conventional chemistry or separation techniques can now take what was a racemic mixture and produce an optically pure compound from it. The success of a biotransformation in making a chiral compound is measured by the enantiomeric excess of the product: the percentage amount by which one of the enantiomers (chiral versions) exceeds the other.

The most commonly used biotransformations involve:

- Acylases (to resolve chemically synthesized amino acids).

- Esterases and lipases (to make a range of esters, lipids, and to resolve fatty acids and alcohols).

- β-lactamases and penicillin acylase (to make penicillins and cephalosporins).

- Peptidases and proteases (to make peptides).

- Steroid transforming enzymes (to make steroid derivatives). These are always used in whole organisms, as many enzymes are involved in each biotransformation.

See **Glycosidases, Lipases, Proteases**.

See also **Chirality**.

Blood disorders

There is a range of diseases of the blood which biotechnologists seek to address. The main ones are:

- Haemophilia. The blood will not clot because the gene for one of the proteins involved in clotting is defective. Several of the blood clotting 'factors' (factor VII, VIII, and IX) have been cloned and are used as biopharmaceuticals to treat these inherited diseases.

- Sickle cell disease, thalassemia (α and β). These diseases are caused by a mutation in the genes for haemoglobin, the red protein in blood cells. Boosting blood production with erythropoietin, replacing the haemoglobin with haemoglobin made in yeast, and ultimately gene therapy to replace the gene have all been suggested and tried on animal models.

- Leukaemia, anaemias. There are a very wide range of disorders in which one of the many types of blood cells are produced in an inappropriate amount. In anaemias, not enough cell, particularly red cells, are produced. Leukaemias are diseases, usually types of cancer, in which one type of white cell is vastly over-produced, usually to the detriment of all other cell types. Leukaemias can be treated by various transplant-type techniques, including transplanting genetically altered bone marrow cells to boost production of the missing type. Production can also be boosted by relevant growth factors and by haemopoietic factors (i.e. factors which boost haematopoiesis, the making of blood in the bone marrow): several of these factors have been made as potential biopharmaceuticals.

Blood products

Originally these were biopharmaceutical products made from human blood, such as the blood clotting factor VIII used to treat haemophiliacs. Such extracted products are usually made by a series of filtrations and solvent extractions. The major 'blood products' in this category are:

• HSA — Human serum albumin. This is the major human blood product by volume, used to produce blood substitutes and extenders for transfusion.

• Human γ globulins. These are antibody preparations, and are used medically to give people an extra high level of antibodies (immunoglobulins) when they might be exposed to specific, unusual diseases.

The term 'blood products' is also used to refer to biopharmaceuticals which act on blood or the cells which make blood. They are also usually made by those cells, but in such tiny amounts that extracting them from blood itself is impractical. So they are made by genetic engineering instead. Among the 'blood products' category of biopharmaceuticals are:

• Thrombolytics: drugs such as tissue plasminogen activator (tPA) produced by Genentech and one of their two products (the other being growth hormone), Streptokinase, Eminase (made by SmithKline Beecham). These dissolve blood clots in the arteries and hence are used as treatments for heart attacks.

• Clotting agents: factors VIII and IX to treat haemophilia, a disease where these proteins are missing. Baxter Healthcare and Miles Inc. are developing recombinant factor VIII.

• Erythropoietin (EPO): this stimulates the bone marrow to make more red blood cells, and is the subject of a fierce patent dispute (*see* **Patents**).

• G-CSF, GM-CSF, etc. (colony stimulating factors): these are cytokines — materials made by the immune cells to regulate the immune system's function (*see* **Cytokines**).

Animal blood products, notably fetal and newborn calf serum, are also used in the biotechnology industry: serums are used as a supplement for the media used to culture a range of mammalian cells.

'Blots'

A range of molecular biological techniques are called 'blots'. They all share a common appearance. At the start, biological molecules are usually present in a jelly-like matrix, often the result of separation by gel electrophoresis The contents of the gel are then transferred onto a porous membrane, often a chemical derivative of paper or a nylon mesh. This technique was traditionally done by allowing liquid to seep through the gel, through the membrane and into a pile of paper towels which acted like blotting paper — the biomolecules travelled with the fluid until they stuck to the membrane. Now electroblotting (which uses an electric field to pull the molecules out of the gel) and vacuum blotting (which uses suction) are also used. Once on the membrane the molecules can be analysed by techniques which would not work in the original gel, such as antibody staining or DNA hybridization (*see* **DNA probes**).

The variations on this theme depend on the molecules:

- Southern blot. Named after Prof. Ed Southern, the gel here is a DNA

Southern Blot

Paper towels (replaced by vacuum pump in some systems)

DNA washed out of gel is trapped here

Membrane (nylon or nitrocellulose)

Electrophoretic gel (contains separated DNA samples)

Wick dipped into reservoir of salt solution

Flow of salt solution

electrophoresis system and so the molecules transferred are DNA molecules.

- Northern blot. Almost the same as the Southern blot, but the molecules are RNA.

- Western blot. Here the molecules are protein, also separated by gel electrophoresis. A common use is to separate proteins according to size by electrophoresis and then identify them by reacting them with an antibody.

- Southwestern blot. This is a variant of the Southern blot used to find protein molecules that stick to DNA molecules.
  (Desperate attempts to get something — anything — named as the Eastern blot have not been generally successful.)

- Dot blot. Here DNA, RNA, or protein are dotted directly onto the membrane support, so that they form discrete spots. Also slot blots, where the sample is applied through slots in a manifold so as to give oval or rectangular blobs of sample, which are easier to quantify.

- Colony blot. Here the molecules (usually DNA) are from colonies of bacteria or yeast growing on a bacteriological plate. A variation (called the plaque lift) can also be used for viruses.

With the advent of PCR there has been a fall in the use of Southern and Northern blots, although these are still widely used.

See also **DNA probes, Gel electrophoresis, Hybridization**.

BST

Bovine Somatotrophin, also Bovine growth hormone. This protein hormone is found naturally in cattle, and is the counterpart of human growth hormone, one of the earliest biopharmaceutical products. It has been cloned, expressed in large amounts and is being marketed by Monsanto as an agricultural product to improve the growth rate and protein : fat ratios in farm cattle, and to improve milk yields.

BST

There are animal welfare concerns about this, and health concerns about the possibility that BST will get into the milk or meat, and hence into people. In particular, the possibility that BST given to improve milk yield will get into the milk given to children has proven a powerful weapon against Monsanto, one of the principle developers of BST for agricultural use. Monsanto has also been accused of reducing cows to unhappy milk-producing machines (*see* **Yuk factor**). The debate has become highly polarized, with contenders on both sides seeing it as a trial case for the application of biotechnology to the agriculture and food industries. The USSR (as it then was), Czechoslovakia, Bulgaria, South Africa, Mexico, and Brazil have approved BST, but in many other countries the debate on its safety is holding up any approval. There is also a debate over whether BST will offer the consumer any advantage, especially in Europe where there is often a surplus of milk over the EC's 'quota' for production. It would, however, allow the same amount of milk to be produced from fewer cows using less food.

Capillary zone electrophoresis

Also simply capillary electrophoresis, this is an up-and-coming technique in many biochemical and biotechnological fields.

Gel electrophoresis is electrophoresis — moving molecules around using electric fields — performed in a polymer material. The polymer does two things: it sieves the molecules by size, and it stabilizes the solution in which the electrophoresis is happening. Without it, any slight vibration or convection would stir all the molecules up, and the systems ability to separate very similar molecules (its resolution) would go down dramatically.

However the separation is a complex result of the molecule's shape, size, charge, and how it interacts with the polymer gel. This complexity can itself reduce the system's resolution.

Electrophoresis without gels has been used. It is called free zone electrophoresis, and uses a flow of water, or sometimes a column of water with the bottom containing more sugar or salt than the top and which is hence stable to stirring. Such density gradients are discussed further in another entry (*see* **Centrifugation**). However the stirring effects can be substantial.

Capillary electrophoresis is free zone electrophoresis in a very fine capillary tube (a tube with an internal diameter of less than 1mm). Here stirring effects undoubtedly occur, but they only stir up volumes of solution less than the tube diameter (i.e. less than 1mm), and so the effect on resolution is very small. The electrophoresis can be 'run' much faster than conventional electrophoresis, where making the molecules go faster means putting a higher voltage across the gel slab, which means that more current flows through the gel, more heat is produced in the gel and ultimately that the biological molecules denature (or the gel tank cracks or bursts into flames). The mass of liquid in a capillary tube is so small that even very high voltages produce tiny currents, and the heat produced can be radiated away from the tube rapidly. So the electrophoresis can either be run very fast or it can be run on a very long capillary tube, so increasing resolution.

There are several commercial systems for performing capillary electrophoresis on biological molecules for research.

Catalytic antibodies

Catalytic antibodies, also called abzymes, are antibodies whose binding sites, instead of passively binding to a target molecule (antigen) catalyse a reaction. Antibodies do not normally possess catalytic activity.

In the 1940s Linus Pauling suggested that an enzyme was simply a protein which bound to and stabilized the transition state of a reaction. By stabilizing the transition state, the enzyme made the reaction from substrate to product more probable, and hence the reaction faster. In the 1960s several workers suggested that an antibody which bound to the transition state of a reaction would catalyse that reaction.

However, it is not possible to isolate the transition state of a reaction. So to raise an antibody against it is impossible. A near approximation is to raise an antibody against an analogue of the transition state. As transition state analogues are often powerful inhibitors of enzymes (as they mimic the transition state to which the enzyme binds), quite a lot are

66

known. Others can be synthesized from considerations of the reaction mechanism. By raising a monoclonal antibody against the transition state analogue, an antibody whose binding site catalysed the reaction concerned can be created. Reaction rate enhancements of 6×10^6 have been reported for some reactions.

Catalytic antibodies can also work through reducing the entropy of reaction, i.e. bringing together two molecules in the right orientation to allow them to react. This can apply to two substrates for a reaction, or a substrate and a cofactor. Catalytic antibodies have been made which catalyse reactions through both these mechanisms. (' Entropy' in this case is chemical entropy, i.e. disorder. Two molecules exactly aligned for a reaction represent a very ordered system — it is much more likely that they will collide in some unsuitable way, or indeed not collide at all. Thus the reaction has a high 'entropy barrier', which the catalytic antibody reduces by making the system more ordered — it brings the two reactants together in the correct way to react.)

As would be expected of a protein catalyst, abzymes are quite specific in the reactions that they catalyse, including selecting only one stereo-isomer from a racemic mix. Reactions catalysed to date include a variety of esterase and peptidase reactions. Abzymes have the advantage that, in principle, a specific abzyme can be created for any reaction. Although an enzyme could also be found for that reaction, finding it can be a major task. The technology of creating an antibody which recognizes a specific small molecule (hapten) is, by contrast, quite simple.

Favoured targets for abzymes include biotransformations, and especially resolution reactions, biosensor applications where the specificity of antibodies can be coupled with the relative ease of detecting enyzme reactions, and pharmaceutical applications. Drugs are particularly favoured targets, as an abzyme which acts as a very specific protease to cleave any protein in the body (such as a viral coat protein or an inflammation-causing cytokine) could be engineered, and would only act on that peptide. Drugs also promise the substantial market size that is needed to justify the very considerable amount of time and money needed to make even simple model abzymes work.

cDNA

cDNA is copy-DNA (or complementary-DNA). It is a DNA copy of an RNA, and is made from the RNA using reverse transcriptase. This is a gene cloning technology. There are two potential reasons for wanting to do this.

Firstly, the DNA gene itself may be unknown. In this case, the cDNA that is a DNA copy of a messenger RNA that codes for a known protein (or for a protein whose activity can be measured, by antibody reaction or because it is an enzyme) may be isolated. Then the DNA gene can be found using the cDNA as a 'probe'.

Secondly, the scientist might not want the 'original' gene. This is especially true if the purpose of cloning the gene is to express it in a bacterium. In this case, the scientist wants a section of DNA that codes for the protein concerned *and nothing else*. She does not want introns, other neighbouring genes, and so on in the gene clone. The cDNA is a much better approximation of this, consisting (in eukaryotes, anyway) of a single mRNA with no introns that codes for a single protein. Often the cDNA can be spliced directly into an expression vector and used to produce the desired protein in bacteria.

cDNAs hit the headlines at the end of 1991 when Craig Venter at the US National Institutes of Health (NIH) filed a patent claiming 337 new cDNA sequences that he had discovered using an automated DNA sequencer (a subsequent patent claimed over 2000 further sequences). Actually the sequences were not complete cDNAs — they were short sections of cDNAs called expressed sequence tags (ESTs), which are long enough to identify a new cDNA uniquely. The NIH's idea is to patent them as Venter produces them, so that, if at any time in the future someone finds a use for those sequences, then the NIH can claim royalties on them. The Medical Research Council in the UK has taken the route of keeping its cDNA sequences generated by large-scale sequencing a secret until the legality and acceptability of patenting cDNAs is established. It seems implusible that the cDNA patent will hold up in its present form: Venter admits that he does not know what these cDNAs do in the cell, and so it is unclear what practical use they can have without substantially more inventive effort.

Cell disruption

Many fermentation processes produce products which are inside microbial cells. Examples are many proteins produced by genetic engineering, enzymes, and large molecules such as the biodegradable plastic poly-hydroxybutyrate (*see* **Biodegradable materials**). It is necessary to break the cells to get these products out. This process is called cell disruption.

The problem is that cells, and particularly bacterial cells, are specifically 'designed' by evolution to be unbreakable. Thus a lot of energy has to be put into breaking them, and there is the risk that that energy will also disrupt the product inside the cell. In general, animal cells are quite easy to break up, plant cells rather hard (as they have a strong wall around them), and yeast and bacterial cells even harder. Methods used are:

- Autolysis. This simply alters the conditions so that the cell digests itself. This is the simplest possible method, but tends to be useless for protein products as the cell digests itself from the inside out, so breaking down the product before the cell wall.

- Enzyme action. This is very effective — the cells are treated which an enzyme with dissolves some key component of their cell wall, which then simply falls to bits. Typical enzymes used are lysozyme (for bacteria), chitinase or glucanase (for yeasts), cellulase (for plant cells).

- Detergents, alkali, osmotic shock (i.e. pure water), plasmolysis (treatment with high concentrations of salt), organic solvents. Any of these treatments will knock holes in the plasma membrane, the thin layer of lipids inside a cell wall which actually holds the cell's contents in (the cell wall, by contrast, is meant to keep the outside world out.) If the product is small enough (as is true of small proteins and of metabolites) or if there is no cell wall (as is true of animal cells) then the product leaks out.

- Freeze-thaw. Freezing and thawing can break up any structure as ice crystals form inside the wet materials of which cells are made.

- Mechanical methods. The most obvious method is to break the cells mechanically. There are many ways of doing this:
  - French press, which forces the cells through a very small hole at high pressure. A large scale version of this is called a Manton Goulin homogenizer.

— Mills, in which the cells are shaken vigorously with abrasives, or metal balls or rods.

— Blenders, traditionally the laboratory uses a blender called the Waring blender (named after a 1930s New York dance band leader who invented or popularized it for making cocktails), but essentially this is a food processor with a powerful motor.

A number of cell disruption techniques produce cells that are lysed, i.e. broken open, but not otherwise disrupted. These cell suspensions can be extremely viscous, mainly because the cells' DNA has not been broken up, and so expands out of the cell to form a dense interlocking network of molecules. Thus many cell lysis treatments include a nuclease treatment step. Nucleases are enzymes which break down nucleic acids, and the object here is to find a very non-specific nuclease that will break down any nucleic acid into very small pieces, ideally only a few bases long. The solution's viscosity then drops dramatically. This also breaks up the RNA in the solution, which is present in much greater mass than the DNA (although it does not contribute to the viscosity problem), and can be a problem in further purification steps if it is not broken down into small fragments.

Cell fusion

The fusion of two cells together results in a new cell which has all the genetic material of the two original cells, and hence is a new type of cell. The ability to fuse different types of cell — from the same species or from different species — has been used widely in biotechnological research. Common methods used include:

• Electroporation (*see* **Electroporation**).

• PEG-mediated fusion. PEG (polyethylene glycol) is a polymer which binds into the lipid membrane of cells and fuses it with any other lipid membranes around. Thus it can mediate the fusion of any cells which are bounded by a lipid membrane (i.e. all animal cells, and plant or bacterial protoplasts).

- Virus-mediated fusion. Some viruses have lipid coats which fuse with the membrane of cells when the virus infects that cell. If the virus fuses with two cells at once, then it effectively joins the cell via a small bridge of membrane. Thus viruses have been used rather like PEG to fuse cells. Indeed, their cell-fusing capabilities were discovered before PEGs were, but PEG is preferred now because it is easier to get hold of and less potentially hazardous.

Cell fusion is used in a variety of techniques. Making monoclonal antibodies relies on making a fusion between lymphocytes and an immortalized cell line. Some plant genetic engineering has used cell fusion to generate hybrid plants, i.e. plants with all the genetic material of two different plant types brought together into one species, by fusing the protoplasts of the two 'parental' species and then regenerating a plant from the result. (This is a difficult trick to achieve.) Polyploid plants, plants with abnormally large numbers of chromosomes, can also be made by fusing cells from the same plant together.

Cell growth

The growth of isolated cells in culture follows a characteristic curve, shown in the figure. The phases of the curve are:

- Lag phase. This occurs when the cells are introduced to their new growth medium, and is the time taken for them to adapt to it. If it is identical to their old medium, then the lag phase can disappear.
- Log phase. This is the main growth phase of the culture, when the cells are growing exponentially. When plotted on a logarithmic scale (on the right of the figure), the log phase shows as a straight line.
- Transition. This is the period (which can be minutes to days) between log phase and the following
- Stationary phase. Here the cells have stopped growing — they have reached the capacity of their growth system to sustain growth.

CELL GROWTH

Cell Growth curves:

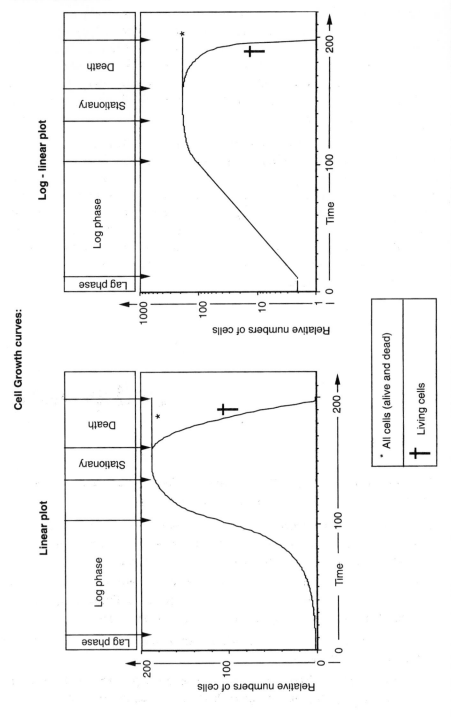

• Death phase. If the cells are not then given fresh medium to start growing again, then they will start to die off. The total mass of cells remains the same (top line) but increasingly few of those cells are alive (bottom line) in the sense that they could start a new growth curve if they were given fresh growth medium.

The length of the different phases varies enormously with different cell types. Thus many common bacteria have a stationary phase lasting only a day or two before death phase starts. By contrast, mammalian nerve cells can last almost indefinitely in culture without dividing. A single mammalian cell isolated from skin or muscle and put into culture medium could take a week to divide — a single *E. coli* cell is unlikely to take more than 10 min to start growth.

The other key idea in cell growth studies is doubling time. This is the time that is needed for the cell population to double in number, and is equal (fairly obviously) to the time an 'average' member of that population takes to go through one complete division cycle. The higher the doubling time, the lower the growth rate of the culture and the longer it will take for an innoculum to reach stationary phase. Doubling time depends on the growth conditions and on the organism being grown — some bacteria (notably *Clostridium perfringens*) can have a doubling time of 10 min in the right culture medium ('growth rate' is sometimes defined as 1/doubling-time). Strictly speaking, the concept of doubling time only applies to organisms growing in log phase, i.e. growing exponentially.

This growth cycle is not the same as the overall life and senescence cycle of primary mammalian cells. Mammalian cells will stop dividing when they have exhausted one of the critical components in their culture medium, or when their surroundings are too crowded for their liking. However if they are separated and put into new medium (a process known as 'splitting' the cells), then healthy cells will start to grow again. Senescence is what occurs when the cells have been split many times so that the total number of divisions they have gone through is 40–60. Then they slowly cease to be able to divide any more, no matter how much new medium they are given.

Cell line

The term 'cell line' is usually applied to mammalian cells cultured *in vitro,* outside their original mammalian body. However it can also be applied to plant cells. A mammalian cell line is a clone of cells, i.e. one which has been derived from one single cell. It is capable of being grown indefinitely, which mammalian cells taken straight from the body are not. Thus the cell has been 'immortalized', i.e. turned from a mortal cell (whose descendants are going to stop dividing after a few dozen divisions) into an immortal one. This can be achieved by transformation of the cell with a virus, with DNA from an oncogene, or by mutagenesis of the cell and subsequent selection of anything that can continue growing.

Cell lines should also be stable, that is they should not change their properties as they are grown. This can be difficult. Unlike normal cells, mammalian cells which have been immortalized often do not pass on their chromosomes very faithfully, and so can loose genes which are not essential to the survival of the cell. These may include genes vital to the biotechnologist, such as the genes making an antibody in a hybridoma cell line. Before a cell clone is described as a cell line, its inventor has to demonstrate that it is stable in this sense.

See also **Immortalization, Strain, Transfection**.

Cell line rights

While a protein can be patented and its ownership is fairly clear, the ownership of a living system is rather more vague. In general, the rulings seem to suggest that any organism which is patentable at all can be patented if it has been manipulated to do something useful, almost no matter how the manipulation is done or how large or small it is. Thus the 'Oncomouse' transgenic mouse has one new gene out of about 100 000, but is still considered a 'new' entity. (For contrast, most mice, and people, probably contain at least half a dozen new, potentially physiologically significant mutations never seen before, as a result of normal genetic change.)

Ownership of the new organism entity usually resides with who made it. It does not reside with the source of material for the new entity: the Moore case in the US (when John Moore claimed that a cell line used in cloning interferon was derived from the hairy cell leukaemia he was treated for in 1978, and hence was at least partly his) ruled that Moore had no rights to his own cell lines. In most countries, people do not have rights over organs removed in surgery: they only have the right to say what happens to their own, whole body on death.

Interestingly, if the Moore decision had gone against Sandoz and Genetic Institute (who now own the cell line), then many other people would have rights to a wide range of cells in research and industry. The descendants of Henrietta Lacks, originator of the HeLa cell line fourty years ago, would now have rights to a substantial part of all molecular biology and a mass of cells which probably greatly exceeds her own weight in life.

Centrifugation

This is one of the more common biochemical techniques, and it crops up very frequently in biotechnological purification schemes. Key terms are:

- Zone vs equilibrium centrifugation. Zone centrifugation places the sample at the top of a tube, puts the tube in the centrifuge, spins it for a limited time, and then takes it out. The product has then 'sunk' a certain way down the tube, separating it from the rest of the sample. If you run the centrifuge too long, everything sinks to the bottom of the tube. Zonal centrifugation separates things essentially according to their size. Equilibrium centrifugation runs the centrifuge until the contents have come to equilibrium, floating, for example, at their buoyant density. Further running will not alter the separation. This is related to the following.

CENTRIFUGATION

- Density gradients. Here the solution in the centrifuge tube is arranged so that it gets more dense towards the bottom of the tube. This is achieved by dissolving something in it: colloidal silica ('Percoll'), to separate live mammalian cells, sucrose to separate bits of cells, caesium chloride to separate nucleic acids, etc. When centrifuged to equilibrium, the sample will be separated according to its density, more dense parts floating further down the tube in more dense solution.

- Density gradient stabilization is also used in centrifugation, as well as in free zone electrophoresis and some other separation techniques. Here again the tube has an increasing density of solution in it, usually sugar solution. This, however, is not done to effect separation. Rather it stabilizes the column of liquid against stirring. If a little bit of solution is stirred out of its 'proper' layer, then it will have a different density to the solution around it, and so will sink back to where it came from.

- Rotors. Most centrifuges consist of a drive unit (which powers it, controls rotation speed, etc.), and a rotor in which the samples are placed and which goes round. The rotor is often removable, and fits in a bowl in the machine. In ultracentifuges (centrifuges capable of tens to hundreds of thousands of times the force of gravity), the bowl is armoured steel to protect the operator should a rotor 'fail' in a run. Legend has it that Svedberg, who developed ultracentrifugation for chemical and biochemical analysis, killed a couple of postdoctoral workers with bits of flying centrifuge.

- Some rotors are 'zonal' or continuous rotors. Liquid is fed in through the middle of them, and bacteria or other particulate matter is centrifuged out to the outside. These are of obvious use in separating microbial cells from their culture medium, but are an expensive way of doing this for large volumes.

Chaperones

These are a type of protein which help other proteins to fold up into their correct three-dimensional structure. A specific, and much-discussed, sub-group of chaperone molecules are the chaperonin proteins. Some proteins will fold up correctly on their own as soon as they are made in the cell, forming a working protein molecule. Some, however, seem to do this very inefficiently, and to need other proteins to make them fold up properly. Strictly, chaperones as a group stimulate any mechanism of getting a protein to fold correctly, by preventing it from folding incorrectly or (the role of the chaperonin proteins) actually catalysing its correct folding.

This is important to the production of 'foreign' proteins in bacteria. If a protein folds inefficiently or very slowly, then it will have a greater chance of aggregating into an insoluble, inactive mass, from which it may be very difficult to recover active protein. If the folding can be speeded up with chaperone proteins, then the amount of *usable* protein that can be recovered from the bacterium (as opposed to the total amount of protein, usable or not) will be greater. Whether the role of chaperones in protein folding can be *used*, as opposed to described, is still an open question.

Chemicals produced by biotechnologists

A number of chemicals are produced commercially by biotechnologists in large amounts (apart from drugs and other specialist materials). Chemicals produced in large amounts by fermentation include:

CHEMICALS PRODUCED BY BIOTECHNOLOGISTS

Chemical	Amount produced world-wide per annum (tonnes)	
Ethanol	75 million	
Acetone	5 million	
Butanol	1 million	
Citric acid	750 000	
Acetic acid	160 000	(much as vinegar)
Glutamate	400 000	
Lysine	80 000	
Other amino acids	20 000	
Nucleosides	5 000	

Chimera

A chimera is an animal which is a mix of several other animals. The Chimera of mythology had a lion's head, a goat's body and a serpent's tail, and breathed fire. More prosaic and realistic chimeras can be made by a range of methods which mix cells from two sources to make an early embryo, which then develops into an animal which has cells derived from two sets of parents.

Chimeras can and have been made by taking the cells from two very early embryos and mixing them together. This can be done at random, or the cells can be selected so that cells which are going to make specific areas of the body can come from one or other of the 'parent' embryos. The techniques of *in vitro* embryology are then used to put the embryo back into a pseudopregnant mother (i.e. a mother animal which has undergone all the hormonal changes necessary to prepare her for pregnancy but who is not carrying any embryos). A sheep/goat chimera was made in this way in the late 1980s (it was called a 'geep'), as was a cow/buffalo chimera. The former attracted such strong public disapproval that the latter was not publicized much (despite combining the resistance to tetse fly of a buffalo with the milk-production ability of good dairy cattle), and further research along this line has been minimal.

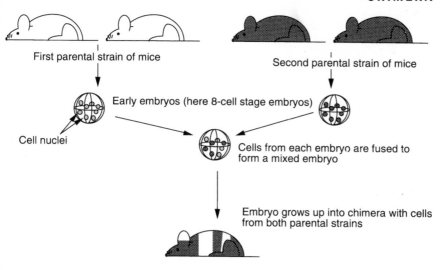

First parental strain of mice

Second parental strain of mice

Early embryos (here 8-cell stage embryos)

Cell nuclei

Cells from each embryo are fused to form a mixed embryo

Embryo grows up into chimera with cells from both parental strains

The animal which has been used in most research into chimeras is the mouse, in which mice of different strains or carrying specific marker genes are used to make chimeras for research. The same method of joining cells from two distinct embryos into one embryo can be used on mice.

Another route is also available, to use cells called embryonic carcinoma cells (EC cells), derived from teratomas (a benign sort of cancer of the reproductive cells). These cells are totipotent, i.e. they can be encouraged to grow into a complete organism. This cannot be done in the test-tube (where the 'embryo' fails to develop for more than a few days) or by implanting the cells in a pseudopregnant mother (where they form a tumour). However, if a few EC cells are mixed with the normal cells of an embryo, then they can be incorporated into that embryo: the resulting mouse has cells derived from the EC cells in many tissues.

If some EC cells get into the reproductive organs, then the mouse can produce offspring derived solely from those EC cells. This is valuable for genetic engineering, as the EC cells can be genetically engineered a lot more easily than mouse eggs can. Engineered cells can then be put into an embryo to form a chimeric animal, some of which is a transgenic animal. This has been proven as a route to generate transgenic mice, but partly because analagous methods for other animals have not been worked out and partly because the embryology is very specialized, this method is used less frequently than microinjection.

See also **Transgenic animals**.

Chimeric/humanized antibodies

A problem with using antibodies in medical therapy is that monoclonal antibodies are foreign proteins, and hence when they are injected the patient will have an immune response to them. This does not matter for a one-shot therapy because the immune response is too slow to have an effect within hours of first encountering a foreign protein. But for longer-term treatment it means that after a few days or weeks the patient will have their own antibodies which bind to and neutralize the immunotherapeutic as soon as it is injected. This is known as the human anti-mouse antibody (HAMA) response (nearly all monoclonal antibodies are made in mice). It is extremely difficult to overcome this by making genuine human monoclonal antibodies as drugs: the technology of monoclonal antibody production works with mouse or rat cells, not human cells.

A way around this is to engineer the antibody so that it 'looks' like a human antibody to the immune system. The species-specific parts of the antibody to which the immune system responds are in the constant regions. Thus by replacing the constant regions of a mouse antibody with those of a human antibody, a protein which binds to an antigen like the original monoclonal antibody but which 'looks' to the human immune system like a human protein can be made. This process is called humanizing the antibody. The fused protein is called a chimaeric antibody.

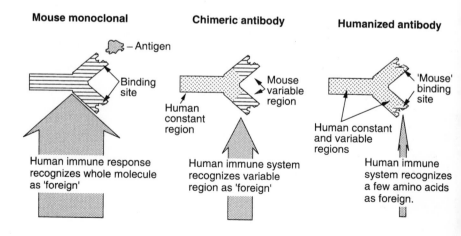

Mouse monoclonal — Antigen — Binding site — Human immune response recognizes whole molecule as 'foreign'

Chimeric antibody — Mouse variable region — Human constant region — Human immune system recognizes variable region as 'foreign'

Humanized antibody — 'Mouse' binding site — Human constant and variable regions — Human immune system recognizes a few amino acids as foreign.

More sophisticated engineering (as not *all* the 'human-specific sites' fall into the constant domains) can be done to produce a humanized antibody. In both cases the antibody gene must be cloned from the mouse hybridoma and engineered *in vitro* before being put back into a bacterium, yeast, or mammalian cell. The ultimate engineering is to take only those parts of the antibody which determine the antibody's binding specificity (the complementarity determining regions — CDRs) and splice them into a totally human antibody.

Engineering antibodies in this way has an added complication. Antibodies consist of two protein chains — heavy and light chains — and so two genes must be engineered into the producer cell to make a finished antibody. While this is possible, and various technical tricks to make it easier have been developed, it would be much easier to handle only one protein chain. This is one advantage of Dabs and SCAs: they are antibody-based proteins which contain only one chain.

See also **Antibody structure, Dabs**.

Chiral synthesis

Chiral synthesis is the production of chiral compounds in only one enantiomer, or 'handedness'. As chiral compounds can be made in two (or more) physical arrangements which are virtually indistinguishable chemically, this is a difficult task for conventional chemistry. Biological systems make just that sort of discrimination all the time, however, and so have great potential for making chiral compounds.

To make chiral compounds in only one enantiomer, there are a range of chemical methods. These include:

• Assymetric catalysis. A catalyst that is itself chiral is used in a key step in the reaction. (Enzymes, of course, are one such catalyst — see below.)

• Chiral chromatography. A racemic mixture of isomers is separated on a chromatographic column that itself is chiral, e.g. has a chiral compound linked to it or is made of a chiral material such as cellulose or protein.

CHIRAL SYNTHESIS

There are several routes to chiral synthesis which use biotechnology. The success of all of them is measured by the enantiomeric excess, the percentage by which one of the enantiomers outweighs the other in the preparation. One hundred per cent enantiomeric excess means that you have an absolutely pure preparation of one optical isomer.

- Biotransformation. This is the synthesis of a compound using enzymes. As most enzymes produce only one enantiomer as a product, they can be used to take symmetric (i.e. not chiral) starting products and produce pure enantiomers from them.

- Bioconversion. This is the same idea, but uses whole organisms to convert one chemical in to another. This may be better than using isolated enzymes if the enzymes concerned are not very stable, or if a number of enzymes are needed to make one conversion. The chiral drug Ephedrine has been traditionally produced by bioconversion.

- Fermentation methods. If you can obtain the chemical from a fermentation culture, either of a microorganism or of a plant or animal cell, then that chemical will almost certainly be produced as one enantiomer. Many amino acids produced for animal feed supplements have been produced traditionally as single optical isomers by fermentation, especially in Japan.

For all these processes, there are two approaches.

- Stereospecific synthesis. This takes two non-chiral starting materials and makes a chiral product from them. This has to be done using some third party to introduce the chirality into the system. This can be a third reagent, or a catalyst: often this chiral catalyst is an enzyme.

- Resolution. This takes a racemic mixture (a 'racemate') of a chiral compound, i.e. a mixture in which all the various enantiomers are present as a mix, and removes one of them. A range of techniques can be used. One isomer may be bound to a material which is itself optically active (such at an optically active HPLC column, or an antibody), but because of their capacity to process only a few milligrams as a time these are usually used as analytical techniques rather than preparative ones. One isomer may be converted into another chemical (which can then be removed by conventional means) using another optically active chemical or, most effectively, an enzyme. The enyzme can either act on the compound you want (turning it into the product, or something more like the product) or on the one you do not want (turning it into something that is easy to remove).

Often, chiral synthesis is not used to make the final chemical itself. Rather it is used to make a precursor to it which is easier to make using the available enzyme systems. This precursor can then be turned into the final chemical using conventional chemistry.

See also **Chirality**.

Chirality

Chirality is the chemical version of 'handedness'. Some molecules have distinct left- and right-hand forms, which, although containing the same atoms tied up in the same way, are not physically the same (just as your hands have the same number of fingers tied to the palm in the same way, but nevertheless are not the same). Such a chemical is called a chiral compound, and the two (or more) forms are called enantiomers (or optical isomers) of each other. Compounds which have two enantiomers are usually divided into l and d, or + and –, or S and R forms, so you have (l)-alanine or (+)-ephidrine. There are complicated rules about these nomenclatures for organic chemists.

Usually there is no chemical difference between the enantiomers of a compound, or indeed between the pure enantiomers and an equal mixture of all of them (called a racemic mixture). The only detectable difference is that they interact with polarized light in slightly different ways. However, nearly all the molecules which make up living systems are chiral. Thus all the amino acids in proteins are l amino acids, not their chemically identical d forms. Because of this, all the chemistry of life is chiral, and so how other chemicals affect life depends on which enantiomer you have; just as it is easier to shake hands left-to-left or right-to-right, not left-to-right (because both hands are chiral: try it), whereas it is as easy to pick up a light briefcase with left or right hands (because, although your hand is chiral, the briefcase handle is not).

This has substantial implications for pharmaceuticals and agrochemicals. Different enantiomers of exactly the same drug can affect biological systems in quite different ways. Thalidomide is a case in point: an effective and safe antinausea agent, the teratogenic side-effects were not due to the drug itself, but to its mirror opposite, the other enantiomer. However the drug was given as a racemic mixture, so the patients got both therapeutic effects and side-effects.

Clearly, as legislative pressures grow for chemicals used in agriculture and medicine to be more specific, there is increasing pressure for any chiral product to be manufactured by these industries as one enantiomer and not as a racemic mixture for these applications. Chiral synthesis is a major aspect of biotransformation and bioconversion technology.

For biopharmaceuticals, of course, chirality is not a worry — as biologically derived proteins, they all have the correct 'handedness' anyway.

Chromatography

Many separation systems used in biochemistry, molecular biology, and biotechnological production are chromatography systems. Chromatography was originally developed as a way of separating pigments from plants by wicking them through paper, an experiment that many schoolchildren do today, and the same basic ideas apply to all chromatographic separations. A small sample is put on one end of a slab or wick of porous material. A solvent is then passed over the sample and up the wick or slab. Depending on whether the molecules in the sample stick to the solid wick or dissolve in the solvent, they either move up the wick or stay put. Most materials do a bit of both, and so move up the wick slowly — the exact speed varies with each component of the sample, and so they are spread out. The pattern of how materials are washed (eluted) off the end of the slab or wick is called the elution profile. This is, in fact, a two-phase separation, and so the two parts of the system are called the mobile phase (the solvent) and the stationary or solid phase (the solid material that the solvent moves over).

There are a lot of variations of chromatography. The more common are:

• Gel chromatography/gel exclusion chromatography/size exclusion chromatography. These sift by molecular size. The chromatographic material has small pores in it, into which small molecules can enter but from which large molecules are excluded. (Different materials have different pore sizes, so that the size limits 'small' and 'large' can be adjusted to suit whatever the scientist is trying to separate.) As the mixture of molecules passes down the column, small molecules diffuse into the pores, where the liquid is stationary, and so spend some of their time standing still. Large molecules cannot enter the pores, and so spend all their time in the moving phase. Thus large molecules move down the column faster than small ones.

• Affinity chromatography. Here a specific molecule is linked to the chromatographic material and molecules separated by their ability to bind to it. If the bound molecule is large and the molecule to be separated small, it is usually called affinity chromatography (*see* **Affinity chromatography**). If the bound molecule is small and the separated molecule large, the method can be called covalent chromatography, although this too is often called affinity chromatography.

• Hydrophobic chromatography. This simply uses a hydrophobic material such as untreated silica as the stationary phase. Molecules stick to it

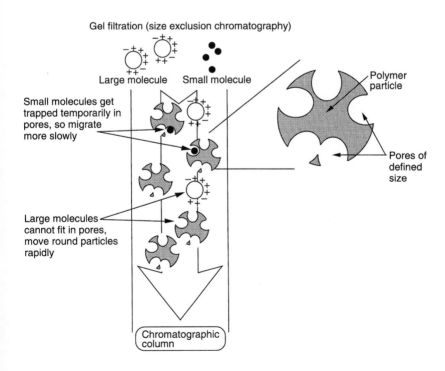

Gel filtration (size exclusion chromatography)

Large molecule   Small molecule

Polymer particle

Small molecules get trapped temporarily in pores, so migrate more slowly

Pores of defined size

Large molecules cannot fit in pores, move round particles rapidly

Chromatographic column

depending on how hydrophobic they are, so it is an effective way of separating many metabolic products.

- Gradient chromatography. Here all the molecules in a sample are bound to the solid support material, and then they are washed off one at a time with an increasing concentration of some solution, often increasing salt, acid, or alkali concentration.

Chromatography also varies according to the physical arrangement of the solid material (the stationary phase):

- Column chromatography. This is the most common by far — the solid phase is packed as small particles into a tube, and the liquid passed over it. Column chromatography methods capable of purifying kilograms of materials at a time have been developed. A variant is high pressure liquid chromatography (HPLC) in which the liquid is pumped slowly over quite a small column at very high pressure. This greatly increases the resolution of the method, i.e. how well it can separate similar materials.

- Paper chromatography. This, essentially the same as the original method, uses a paper wick as the solid phase. This is not as limited as it may sound, as paper is a complex material, and papers with widely varying properties are available.

- Thin layer chromatography (TLC). Here the stationary phase is a thin layer of treated silica coated onto a glass plate.

Lastly, there are the different materials that can make up the mobile and solid phase. In general the mobile phase is water, or some watery solution — this is because nearly all the materials that biotechnologists are interested in are soluble to various degrees in water, and some like proteins are not stable in any other solvent. The solid phase gives greater flexibility:

- Polysaccharides. Much loved by biochemists are polysaccharides such as cellulose (both as a granular material and as paper), Sepharose and Sephadex (tradenames for crosslinked polysacchardide beads), and agarose. They are used for gel chromatography and for affinity methods.

- Synthetic polymers. Coming into increasing favour are synthetic polymers such as polystyrene, PMMA (Perspex), and Teflon, because they are easier to form into small, rigid, uniform spheres and are chemically more robust. Polyacrylamide is also used.

- Silica. Chemically modified silica, especially silicas with chemically modified surfaces and silica materials with a porous structure (CPG — controlled pore glass) are used in many applications. For applications involving substantial pressures such as HPLC (in which polysaccharide beads are likely to squash) silica is very useful.

Generally, chromatographic methods are used to separate several different chemicals from a mixture at once.

Clean room

A clean room is a room which meets specific standards of cleanliness, especially with respect to what may go into and out of it and what the concentration of particles in the air might be. Clean rooms are central to pharmaceutical manufacturing, as it is by producing, formulating, and packaging drugs under rigorously sterile conditions that the sterility of the drug is assured. The same cleanliness conditions apply to a lesser extent to other biotechnological products, and also can apply to the research and development phase of recombinant DNA or plant or animal cloning work, where here the objective is to stop contamination of experiments.

Clean rooms are classified in the US according to US Federal Standard 209D. Roughly, clean rooms are classed by a number, which is the number of particles of more than half a micrometer diameter which are allowed per cubic foot of air. Thus a class 100 clean room would have around 100 0.5 μM particles ft^{-3}. (The exact number varies slightly from this). At the moment, class 100 rooms are the highest category of cleanliness required by the pharmaceutical industry. Other countries have different rating systems (notably, most are based on SI units), but the level of air purity required is similar.

Clean rooms are kept clean by a variety of methods. Air going into the room is filtered so as to keep out even very small particles: ultra-clean rooms have several layers of filtering. Walls, floor, and ceiling are usually painted with materials designed not to hold any dust (and, obviously, not to peel or flake). People going into the clean room have to wear hats and overshoes, as hair and shoes are the most particle-laden parts of the worker, as well as the usual laboratory coat. For less rigorously clean areas there may be a sticky mat just inside the door, which pulls loose dirt off the bottom of the shoes of everyone who walks in.

To provide greater cleanliness within a clean room, laminar flow hood are provided. These are benches either made of, or surrounded by, open meshwork, and covered by a hoods. Air flows up past the worksurface and into the hood, where it is filtered before being returned to the worksurface. Thus all the air passing over the workplace is in a separate flow from the air in the room, and has been extra cleaned.

Clean rooms use much the same air filtration technology as containment laboratories, but for a different purpose. Containment laboratories are meant to to keep possibly hazardous material *in* the laboratory, rather than external contamination *out*.

See also **Physical containment**.

Cleaning-in-place

This is the cleaning and sterilization of a bioreactor system without dismantling it, so that the parts are cleaned as a whole: it also called '*in situ* sterilization'. This is a much easier operation to perform than cleaning and sterilizing all the components separately and then trying to assemble them under sterile conditions, or having separate cleaning and sterilizing operations. However it requires some specialist techniques and equipment.

In particular, the bioreactor machinery must be designed so that there are no 'dead legs' (i.e. pipes that are blocked at one end), crevices, or 'shadowed' areas (i.e. areas where the bulk of some other piece of the apparatus prevents fluid from flowing) into which cleaning fluid could not flow. It is also useful for the equipment to be so designed that bits of it can be cleaned while the rest remains in operation.

Clone

A clone is a collection of genetically identical individuals, which have been derived from a single parent. They turn up in molecular biology and biotechnology in many contexts.

- Clones of organisms. Clones of plants and some animals have been developed using many techniques. Members of a clone show less variability than collections of the same organisms bred by sexual methods, and cloning may provide a method for the rapid multiplication of some desirable individual without having to wait for breeding cycles. Cloning plants usually involves plant cell culture. A plant is broken up into very small pieces, even single cells. These are grown into larger masses in culture, and then these masses (calluses) are induced to differentiate into the tissues of a plant. This route is particularly useful for propagating plants with long life-cycles, such as trees.

- Cloning animals is more difficult, and relies on some manipulation of their normal reproductive cycle. Mammals may be cloned by splitting the very early embryo into several smaller clusters of cells and growing each as a separate embryo: usually, no more than eight individuals may be produced in this way. Fish and frogs may be cloned in larger numbers.

- Gene clone. This means a collection of organisms (usually bacteria) which all contain the same piece of recombinant DNA. By extension, it means the piece of DNA they contain (*see* **Recombinant DNA**).

- Cell cloning. Some biotechnology methods produce a collection of single cells which are genetically different. Producing hybridomas is an example: the fusion step produces a large number of different fused cells. These variants are then 'cloned', i.e. separated out, and individual cells grown up to produce a clone of cells.

Clubs

Many countries have set up various collaborative efforts between companies, and between industry and academe to encourage information transfer in biotechnology. Generally their function is to encourage research with no immediate commercial application. Usually they have government funding to support research initiated or funded by industry. Among the supporting institutions are:

- US State biotechnology centers. A wide range of types of institute which sponsor biotechnology research and provide financial and, sometimes, technical aid and advice for setting up biotechnology research groups or companies.

- SERC (Science and Engineering Research Council) and DTI (Department of Trade and Industry), UK. The bodies have set up several collaborative endeavours such as the LINK schemes and 'Clubs' in protein engineering, sensor technology etc. to match industrial funding for research with government money, and to encourage intercompany collaboration.

- MITI (Ministry of International Trade and Industry), Japan. Known for its support of the Japanese semiconductor industry, MITI has set up the Protein Engineering Research Institute, a 14-company research consortium funded with approximately US$100 million of government money.

Coenzyme

The term cofactor is used almost interchangeably with coenzyme in most contexts. A coenzyme is a molecule which is needed by an enzyme to work, is part of the chemical mechanism of the enzyme, but which is not produced for its own sake. Rather it acts as a shuttle molecule, taking groups between one enzyme and another. Thus it does not act as a catalyst in itself, but rather acts as a catalyst of the transfer of atoms and molecules between enzymes.

The most biochemically common coenzyme set is the NAD set. These molecules shuttle hydrogen atoms around the cell. There are two flavours (NAD and NADP), and they come as hydrogenated (reduced) or non-hydrogenated (oxidized) molecules — NAD or NADP = oxidized, NADH or NADPH = reduced.

Many cofactors and coenzymes are derived from vitamins. Thus NAD is derived from nicotonic acid.

Some coenzymes are tightly, even covalently linked to their enzymes — it is these that are often called cofactors. An example is the FAD (flavine adenine dinucleotide) moiety that is needed by the common diagnostic enzyme glucose oxidase. If the FAD is removed the enzyme does not work. Such a cofactor-less enzyme is called an apoenzyme. It contains all the protein of the intact, functional enzyme (the holoenzyme), but does not catalyse its reaction.

Coenzymes are also of concern to biotechnology in two other fields. Firstly they are often inconveniently complicated and expensive molecules to make and store, and so research is looking for synthetic alternatives. Secondly, some abzymes have been made which use coenzymes to catalyse reactions.

See also **Biomimetic, Catalytic antibodies.**

Computational chemistry

This is blanket term for using computers to predict or analyse the properties of molecules (as opposed to using computers simply to draw them, which is molecular graphics). Calculating the properties of molecules from first principles, which would be ideal, is impossible for practical purposes. Thus computational chemistry uses the known properties of chemicals to calculate the properties of similar molecules partly from empirical rules ('heuristics'), partly from 'rigorous' calculation.

One of the main areas of interest is predicting how proteins will fold up. In principle this should be predictable from their amino acid sequence, but this is not yet achievable. So there are a series of 'half-way houses'. The most 'rigorous' method is to model the peptide chain as a series of links with known charge, hydrophobicity (i.e. propensity for *not* dissolving in water), etc., and see how they all interact with each other. In principle, this leads to a prediction that the protein will fall into a compact, stable structure. At the other extreme, one looks for a similar protein whose structure is known from X-ray crystal studies and tries to fit the amino acid sequence of the protein under investigation to that, known structure. Half-way houses involve taking that 'fitted' structure and then 'optimizing' it using chemical calculations. Another approach is to search a database of structures (such as the Brookhaven database, compiled by the National Laboratory at Brookhaven, Connecticut in the US) for *bits* of proteins which have the same amino acid sequence as *bits* or your protein, and then compile a final structure from those bits. There are also algorithms for looking for short sections of amino acid sequence which have been found to form particular bits of proteins: these bits can then be compiled into a final structure.

The reason for doing this is to be able to predict the structural and functional properties of a protein. This is particularly important for drug discovery programmes, where the properties of a protein can be used to predict what compounds will bind to it, and hence modify its behaviour in a medically useful way.

Although computational chemistry is distinct from molecular graphics, the two have a close link. The results of computational chemical calculations are often displayed as computer pictures of molecules. And one route around the enormous difficulty of computational chemistry is to use the human brain as a computer to analyse molecular patterns displayed on a screen.

See also **Molecular graphics**.

Concentration

Biological products are usually produced in rather low concentrations, by fermentation methods or by extraction from animal or plant tissues. In order to keep the cost of purifying these materials down, it is useful to reduce the volume, i.e. to increase the concentration, as early as possible in the downstream processing stages of a biotechnological process. Many concentration methods also purify the product to some extent as well. The very best concentrate and purify in one step, but that is rarely possible.

Methods used in concentration are based on:

• The size of the molecules. In this category come various filter methods and reverse osmosis. In reverse osmosis, the sample is placed on one side of a semipermeable membrane, that is, one which will let through water but not other materials. A very high pressure then pushes the water through the membrane, giving water on one side and much concentrated product on the other. This can be a way of purifying water, too — it is sometimes used to make drinking water from seawater. It is the reverse of osmosis, the process by which water moves from one side of a semipermeable membrane to the other side if the concentration of soluble material is greater on the other side. Ultrafiltration is a similar technique. Here the molecules are filtered through a membrane with pores of molecular size. Large molecules stay on the sample side, while water and small molecules including salt pass through. Again, considerable pressure is usually needed to make this happen.

• The charge of the molecule. This usually means ion exchange methods. Here a polymer is synthesized with a charge on it: usually, this is a polymer with charged side-groups. Molecules with the opposite charge to the one on the polymer will stick to the polymer. A large volume of dilute product can be washed over a small amount of ion-exchange polymer (or resin, as they are usually called), and the product concentrated on it. Product can be washed off again by washing with acid or alkali, or sometimes with concentrated salts.

• The solubility or volatility of the molecule. The former includes counter-current extraction methods, in which two immiscible liquids flow past each other, and the material you want is successively exchanged from one liquid to the other. The latter are essentially variations on distillation, which is not usually applicable to large charged biological molecules.

If the product is not a molecule but rather is whole cells, then methods based on the relatively large size of the cells can be used. These include:

• Sedimentation. This simply collects the cells by allowing them to fall out of the culture medium. It works well for large fungal mycelia or animal or plant cells, as these can settle out in a matter of hours. However some bacteria could take days or weeks, as they are very small, and ones that can swim by themselves would never settle out at all. Other methods can be used here, or they can be centrifuged to speed up separation: however centrifuging large volumes of liquid can be expensive.

- Flocculation (making the cells clump together and then letting them settle out as a visible precipitate). This is used quite extensively in the sewage industry.

- Flotation (as cells can get stuck onto the walls of bubbles, and so be carried to the top of the liquid to be collected there as foam.) This is a well-known technique from the mining industry.

Cross-flow filtration

This is a commonly used method for filtering the sorts of dense and thick fluids which have to be filtered in biotechnological separations in order to concentrate some material. If one tries to filter (say) soup through a standard micropore filter in order to concentrate the particulate material, the pores will rapidly block up, and filtration will be brought to a halt. Cross-flow filtering does not seek to filter the liquid through the filter directly. Rather, it flows the liquid across the filter, allowing the carrier fluid to flow through. After it has passed over, the top (unfiltered) phase is more concentrated, and some of the fluid phase has passed through — meanwhile, the filter is not blocked up.

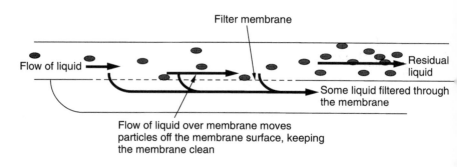

Filter membrane

Flow of liquid →

Residual liquid

Some liquid filtered through the membrane

Flow of liquid over membrane moves particles off the membrane surface, keeping the membrane clean

Cryopreservation

This is preservation of things by keeping them cold. There are several variations of relevance to biotechnology.

- Freezing. This is the most obvious route. Putting something in the fridge or freezer is fine for many biological materials, but not all, as the process of freezing sometimes destroys what you are trying to preserve. This is especially true of cells.

- Freezing in mixed solvents. To prevent damage to cells on freezing, they are often frozen in a mixture of a watery material (their usual growth medium) and another liquid which mixes with water. The other liquid prevents the water from forming ice crystals, which would otherwise disrupt the cells. Glycerol is a favourite for bacteria, dimethyl sulphoxide (DMSO) for animal cells.

- Bacterial cells preserved in this way can be kept in a conventional freezer, but animal cells need to be stored at liquid nitrogen temperatures if they are to survive more than a few weeks. This is often called storing them in liquid nitrogen vapour phase, as the tubes of cells are kept in a flask of liquid nitrogen *above* the nitrogen itself, so that they are not actually immersed in the liquid, only exposed to its vapour. Apart from anything else, this stops the tubes filling up with liquid nitrogen and then exploding when you warm them up.

- Antifreeze proteins. There are proteins which prevent ice crystals forming found in some arctic fish. In principle they could be used to replace the glycerol or DMSO (which are somewhat toxic), but this is rarely done in practice.

- Freeze-drying. This is not a cryopreservation method really, as the dried sample is not stored cold, but is often classed as one (*see* **Freeze drying**).

Culture collections

Many countries and institutions have set up places where samples of microorganisms and cell lines are stored. They may also be called strain depositories or type culture collections, the latter because they are where the 'type specimens' (i.e. the definitive specimens which describe that 'type' of organism) are kept. They have a triple function. They are a 'bank' for valuable microorganisms (against the risk that your laboratory burns down). They are a centre from which other people can get samples of your organism (if you want them to) without bothering you. And they are somewhere where you can deposit an organism to prove you own it — a sort of biological patent office. Some patenting systems insist that you deposit a sample of any organism mentioned in a patent, which cannot be created easily by someone else, at a recognized depository so that, if there is a dispute later, there is some way of proving what your original organism looked like.

The best known depository is the American Type Culture Collection (ATCC), which collects all types or microorganism and cell line. ATCC is also the World Health Organization (WHO) international reference collection. There are a variety of other general depositories in other countries, some of which specialize in fungi, bacteria, or animal cells. There are also industry-specific depositories for the dairy industry, marine organisms, pathogens, etc. This can be confusing to someone trying to find a specific organism, so there are a number of centres and databases to help track down organisms. Europe has a culture collection of purely mammalian cells — the European Central Animal Cell Culture facility (ECACC) at Porton Down, UK.

Cyclodextrins

These are cyclic carbohydrates made of six, seven or eight glucose molecules joined in a ring, to form α-, β- and γ-cylodextrin, respectively. They are synthetic molecules, made by biotransformation. The cyclodextrins all form cylindrical molecules with their water-soluble groups on the outside of the molecule and a relatively non-polar hole down the middle. This hole can accommodate another molecule, known as the 'guest' molecule. This allows them to be used in a wide range of applications, including improving the solubility of drugs and biopharmaceuticals, and selectively binding materials which 'fit' into the central hole in affinity purification and chromatography methods (*see* **Affinity chromatography**).

Natural cyclodextrins are not used extensively in drug applications, because they are not very soluble and are rather toxic to injection. However they may be modified by adding alkyl or hydroxyalkyl groups onto the hydroxyls of the natural cyclodextrin, which reduces toxicity and can enhance solubility.

Cytokines

Cytokines are materials which stimulate cell migration, usually towards the source of the cytokines. Cytokines are studied in mammals because they are important to many processes which involve cells moving about, such as inflammation and development. Understanding them, and then isolating them and producing large amounts for therapeutic uses, is a major research target of many 'genetic engineering' and pharmaceutical companies.

Most well characterized are the cytokines which act on the cells of the immune system, attracting them to sites of damage or infection where they can kill invading cells and, as a side-effect, produce inflammation, shock and even death. So well understood are the immune system cytokines (compared to other cell mobility enhancers) that 'cytokine' usually refers exclusively to cytokines which act on lymphocytes and macrophages. Cytokines are also involved in the body's control of how many blood cells are made in the bone marrow, and so are of general interest as potential stimulators of blood production (haematopoeisis). A review of cytokines is beyond this book, but the ones known to date include:

Interleukins. There are eight known (IL-1–IL-8). IL-2 has been used as a booster of the immune system in cancer and infectious disease therapy: it stimulates T cells to proliferate. IL-1 has several effects with the overall effect of stimulating the production of blood cells by the bone marrow, as well as stimulating non-immune cells to produce other cytokines. IL-4 is linked to the allergic response (IgE-mediated immunity), and so agents which affect IL-4 response have potential for modulating allergies.

CD antigens. Many of the CD antigens which allow scientists to distinguish different types of lymphocyte are interleukin receptors: that is, they are the proteins which interleukins bind to, and through which Interleukins have their effect on a cell. CD (it stands for cluster differentiation) antigens turn up in a variety of contexts, most notoriously CD4 as the protein that the AIDS virus uses to bind to its target cells.

Colony stimulating factors (CSF). There are three varieties: G-CSF, M-CSF, and GM-CSF, which stimulate granulocytes, macrophages, or both (respectively). They stimulate the differentiation of some types of white cell. Ten companies are trying CSFs as biopharmaceuticals.

Interferons (IFN). Well known as being one of the first proteins to be produced by the new biotechnology of the late 1970s, and touted as the wonder-cure for everything, there are actually three classes of these cytokines. They are now consistently called interferons α, β and γ. IFN-γ is a potent stimulator of the activity of macrophages, encouraging them to kill tumour cells and intracellular parasites. Interferon A (from Biogen) has recently been approved as a treatment of hepatitis C by the FDA. Bovine interferon has also been shown to help improve the pregnancy rate in sheep, because it increases the process of 'maternal recognition' by which the ewe's immune system learns that the developing fetus should not be rejected. This unusual use of a cytokine could be as widespread as medical uses.

Tissue necrosis factor (TNF). Slows cell growth and kills some cancer cells and cell lines. It is therefore a hot candidate for an anti-cancer drug, and as the 'toxin' part of an immunotoxin. It is also involved in the cell destruction which can occur in some inflammations, so finding ways to block TNF's action is also a hot pharmaceutical topic.

Several companies are developing genetically engineered cytokine preparations for drug use: Genentech (γ–interferon), Cetus/Chiron (IL-2), Immunex (GM-CSF).

Dabs

These are antibodies in which there is only one protein chain derived from only one of the 'domains' of the antibody structure, and are hence called single domain antibodies or Dabs. Greg Winters at Cambridge, UK, has shown that, for some antibodies, half of the antibody molecule will bind to its target antigen almost as well as the whole molecule. Usually the binding site of an antibody consists of two protein chains.

The potential advantage of dabs is that they can be made easily by bacteria or yeasts. Whole antibodies have two protein chains, and therefore need engineering with two genes. Gene cloning vector systems exist to do this, but it is relatively tedious. Dabs offer a way to clone antibody-like molecules into bacteria, and hence to be able to screen millions of anti-bodies much more easily than it is possible to screen monoclonal antibodies.

Related ideas are single-chain antigen binding technology (SCA), patented by Genex, biosynthetic antibody binding sites (BABS) invented by Creative Biomolecules, and minimum recognition units (MRUs, or complementarity determining regions — CDRs), which is a more general description of the smallest part of an antibody you need for it to bind to its target. SCAs are antibody binding domains in which the two chains are linked by a short peptide, so they can be produced from one gene. This makes them much easier to produce in bacteria from recombinant DNA, as there is no need for the two chains of the normal antibody structure to be made separately and then assembled within the cell.

In most of these antibody-derived protein systems, the idea is to use the immune system to generate a 'random' binding site which the genetic engineer then builds into a molecule and which is more convenient to use than an antibody. Thus they are really specific examples of the idea of Darwinian cloning.

See also **Antibody structure, Darwinian cloning**.

Darwinian cloning

This means selecting a clone from a large number of essentially random starting points, rather than isolating a natural gene or making a carefully designed artificial one. From this mixture you select, by whatever means at hand, those molecules which look more like those you want than the rest. (How you select them depends, of course, on what sort of molecules you want.) You mutate these to generate a new set of variants, and re-select, make more variants, and so on until you have the molecule you require.

There are several classes of catalytic molecule suitable for this.

- Catalytic antibodies (*see* **Catalytic antibodies**). Indeed all antibodies are evolved in this way: the body does the randomizing and selection procedures in the immune system.

- Random proteins. In principle, one could clone a totally random bit of DNA in an expression vector, measure enzyme activity, make alterations in the DNA of the clones which show the best activity by random mutagenesis, select again, and so on. However this is tedious, as there is usually a fairly complex procedure for converting a piece of DNA into expressing clones of yeast or bacteria, and then assaying the result. (The protein need not be a catalyst: it could be a peptide which is bound to a receptor protein, or even a molecule with interesting structural properties.)

- A variant of random proteins is 'fusion phage' technology. Here the random protein is part of the coat protein of a **bacteriophage**. A large number of bacteriophages are made, each with a different random protein spliced into them. When the phages infect a host cell, they produce infectious virus particles with the random protein scattered over the outside. This can then be captured using an antibody or assayed for enzyme activity. The 'winning' phage is then simply grown up in bulk to provide lots of the desired protein.

- Antisense. The word 'aptamer' has been coined for antisense RNAs and DNAs which have been selected by Darwinian cloning to bind very tightly to a particular gene or RNA. The start-point here is a random chain of bases, which is bound to the target molecule. Those that do not bind, or only bind weakly, are simply washed away and discarded. The few molecules (out of billions) that are left are then detached and amplified using PCR.

● Catalytic RNA. RNA can also be selected in this way, but with the added advantage that RNA can be a catalyst in its own right. This sort of Darwinian selection has been done to make RNAs which will bind specific low molecular weight chemicals very tightly. The next step is to find one which binds a transition state analogue for a reaction which would, plausibly, make a new catalytic RNA.

The advantage of Darwinian systems is that they select a new catalyst from a vast number of possibilities. There are more possible 100-amino-acid proteins than there are electrons in the universe, so screening them all is clearly impractical. However this approach edges up to the desired catalyst one step at a time. If the catalyst you want has not been found in nature, then this could be an approach to getting it. A company — Affymax — has been founded specifically to take advantage of such technologies. Of course, many other groups are using similar methods, all of which are still experimental.

See also **Antisense, Catalytic antibodies**.

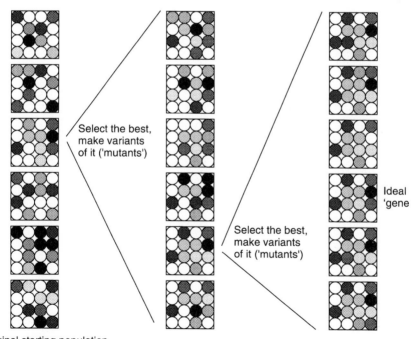

Select the best, make variants of it ('mutants')

Select the best, make variants of it ('mutants')

Ideal 'gene

Original starting population of 'genes': generated at random

DELFIA

This is a trade name for delayed fluorescence immunoassay, marketed by Pharmacia. It is one application of a type of fluorescence detection called time resolved fluorescence. The problem with fluorescence as a method of detection is that it is impossible to distinguish between the fluorescence of the 'marker' molecule (that one wants to detect) and the fluorescence of everything else in the sample, including the sample holder (that one does do not want to detect). A solution is to use a fluorescent material that has a long 'fluorescence half-life', i.e. one that goes on fluorescing for a long time after the exciting light source has been turned off. One looks at the fluorescence after the exciting light has been turned off.

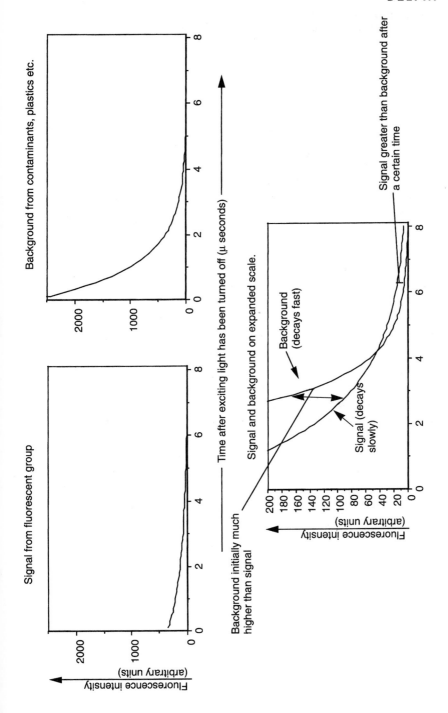

Background from contaminants, plastics etc.

Signal from fluorescent group

Time after exciting light has been turned off (μ seconds)

Signal and background on expanded scale.

Background (decays fast)

Signal (decays slowly)

Signal greater than background after a certain time

Background initially much higher than signal

Deliberate release

This is putting something into the outside world ('the environment'), usually meaning putting a genetically manipulated organism into field trials. Such organizers are often called GMOs — genetically manipulated organisms, or sometimes GMMOs — genetically manipulated micro-organisms. A wide range of such trials have been suggested, and some carried out — the first was probably the trial of a genetically engineered frost-proofing bacterial strain in California in 1986. By the end of 1989 there had been 140 deliberate release experiments in the US ,and about half that number in Europe.

A wide range of social, scientific, and political pressure groups support and oppose such trials, on the basis that the organisms may be, or are known to be, dangerous. The biotechnology industry considers the fears to be widely exaggerated, and complains that every time they take a precaution to allay those fears, the opponents of the release experiments take it as proof that the organisms concerned really *were* dangerous after all.

Greenhouse trials are the natural follow-on from laboratory trials, and then, for organisms involved in agricultural applications, deliberate release experiments. Laboratories can include a range of barriers to stop genetically engineered organisms getting out: negative pressure rooms, sterilization procedures, and the genetic engineering of the organisms so that they cannot survive in the outside world. But, of necessity, none of these can apply to release into the outside world. Here affected fields, animals, soil etc. is kept isolated from surrounding farms, and affected material is destroyed after the trial (except for some Australian pigs, which accidently got to market as human food in 1988).

See also **Regulation of organism release**.

Desulphurization

One specific area of environmental biotechnology which has been attracting attention is desulphurization of oil and coal. Sulphur residues in fuels end up as sulphur dioxide when the fuel is burned, causing acid rain. However sulphur-containing fuels are often cheaper than 'clean' fuels. As a rough guide, a 'high sulphur' coal would contain 6 per cent sulphur (mostly as pyrites), and cost US$50–100 t^{-1} less than a 'low sulphur' coal containing 1 per cent sulphur or less. Thus there is an economic motive for removing the sulphur from oil and coal.

The same types of bacteria which are used in microbial mining can be used for desulphurization of coal. They oxidize sulphides (which are insoluble) into sulphates (which are soluble). The sulphates can then be washed away, with the bacteria. This does not work well on lump coal, as the bacteria cannot get inside the lumps fast enough to be economic, but can be effective for treating pulverized coal such as that used in electricity generating stations.

Crude oil also contains significant sulphur — between 0.1 per cent (from the Far East) to up to 3 per cent (for some Middle East oils). Usually oil has its sulphur removed by Hydrodesulphurization, a physico-chemical technique, but work on using bacteria to remove the sulphur shows that it has some potential.

Disulphide bond

This is a form of chemical bond in proteins much talked about by biotechnologists because of its role in stabilizing their three-dimensional structure, and hence the normal function, of proteins. It forms when two cysteine amino acids in the protein react to form one cystine residue. They link via their sulphur atoms, which therefore form a bridge of two sulphurs between distant parts of the peptide chain that fold close to each other in space. Once linked in that way, the chain is locked into that fold, as to unfold again would mean breaking the covalent bond.

Biotechnologists have used genetic engineering approaches to make proteins more stable, by inserting pairs of cysteine residues into the chain at places that are next to each other when the chain folds up. They will then (the idea goes) link up to form a disulphide bridge and so hold the protein more tightly in its native conformation.

DNA amplification

This is the use of enzymes to take a piece of DNA and multiply it in a test-tube into many thousands of millions of copies. It is of enormous potential use in detecting when specific genes are there without using radioisotopes to detect them. The best-known and by far the most commonly used is the polymerase chain reaction (PCR) system of Cetus.

Other systems announced or being developed include the following. (The author will not attempt to describe them all in detail here!)

- Ligase chain reaction (LCR). Uses DNA ligase, the enzyme which joins two DNA molecules together, to link two oligonucleotides together if a target DNA is present.
- Nucleic acids sequence-dependent amplification. This creates a new molecule of the DNA joined onto a promoter for RNA polymerase. The amplification cycle occurs when the RNA polymerase copies this DNA onto RNA, which is then turned back into DNA by reverse transcriptase. The advantages are that this all occurs at one temperature, and that the RNA polymerase creates many RNA molecules from one DNA molecule, so it has the potential for being very efficient.
- There is also an RNA-based system, the Q-β system of Gene-Trak. The RNA of a small virus — Q-β — is duplicated by the RNA polymerase enzyme which the Q-β virus carries. Add one molecule of Q-β RNA to a tube of Q-β replicase and the right chemicals, and the tube fills up with Q-β RNA. The Q-β replicase amplification system uses the enzyme to replicate RNAs which are related to the original RNA but

have a probe sequence in them. Unlike the other systems above (which are target amplification systems), this is a probe amplification system.

All these systems are being developed to be used in medical diagnostics as well as research. All suffer to a greater or lesser degree from the problems of their extreme sensitivity to contamination.

See **PCR**.

DNA fingerprinting

DNA (or genetic) fingerprinting, or profiling, is a way of making a unique pattern from the DNA of an individual, which can then be used to distinguish that individual from another. Most DNA fingerprinting systems rely on DNA probes, short pieces of DNA which hybridize to the genes from an individual to identify specific pieces of DNA within the total collection of DNA. The original DNA fingerprinting probes, discovered by Prof. Alec Jeffreys, used 'minisatellite' DNA, DNA which hybridized to short runs of bases called minisatellites which vary greatly between individuals. Because there are 50–100 of each type of minisatellite in each person, the chances that any two individuals will have the same pattern of *all* the minisatellites is miniscule unless they are related.

Different DNA fingerprinting systems use different probes. It is also possible to create 'single locus probes'. Whereas the normal DNA finger-printing probes create a pattern like an irregular ladder to be compared between individuals, single locus probes detect just one DNA sequence — one rung on the ladder. This makes comparison between two individuals easier.

PCR has been used for DNA fingerprinting in two ways. Firstly, PCR can be used to amplify tiny amounts of DNA into amounts that are large enough to detect using conventional PCR techniques. Secondly, PCR can be used to find random pieces of DNA which happen to be highly variable between individuals. This is called RAPD — random amplifica-tion of polymorphic DNA

DNA fingerprinting has been used extensively as evidence in paternity,

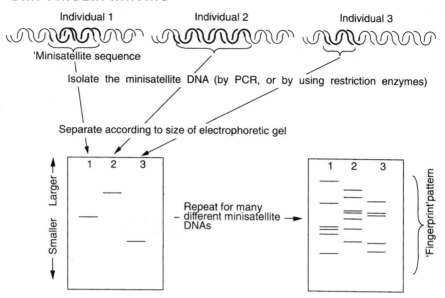

rape, and murder cases to identify individuals. Until 1989 it was considered unassailable evidence, but since then several cases have raised quest-
ions about poorly collected or poorly analysed DNA fingerprinting data,
starting with the case of State vs castro in New York, where supposedly
watertight DNA fingerprint evidence was refuted on technical grounds by
the defence. This led both to a better understanding of the strengths and
weaknesses of DNA fingerprinting and to better quality control in DNA
fingerprinting laboratories.

DNA probes

As well as being used as the genetic material to 'program' cells to do
things, DNA can be used as a reagent in its own right. DNA used in this
way is almost always used as a DNA probe, also called a hybridization
probe. One strand of the DNA double helix is used to bind to a target
strand of DNA. If the base sequences are complementary (i.e. Adenine

matches with thymidine, guanine with cytosine), then the two strands will form a double helix. If they are not complementary, then no helix will form. Thus the DNA probe can be used as a reagent to detect when a specific DNA sequence is present among a mixture of sequences. This process of getting a DNA probe to bind to a target sequence is called hybridization, and can be used to detect DNA or RNA.

DNA probes have been used in genetic research for over thirty years, but only became common when DNA cloning enabled pure DNA probes to be derived from just one gene. DNA probes are still the standard method for finding a DNA sequence among a mixture, often allied with the '**blot**' technology of analysing complex mixtures of DNA molecules.

DNA probes are particularly used in medical genetics as a way of finding whether a person carries a particular gene or not (although in this application they are gradually being replaced by PCR-based techniques). They also have potential for use in detecting pathogenic bacteria, although this has not been realized as fast as was expected in the early 1980s. Probes are also the basis of DNA fingerprinting (*see* **DNA fingerprinting**).

One common use for DNA probes is to find a gene similar to one already possessed. Thus if I have a clone of a gene that performs a useful function on one organism, I can use the DNA from that clone to identify the similar ('homologous') gene in a range of related organisms. (Actually, purists insist that 'homologous' has a different definition, but few technologists are purists.) This is in contrast to heterologous probing, where the DNA probe is used to find a gene that is only similar, not virtually identical, to the one from which the probe was made. This can be useful for cloning, say, heat-resistant enzymes from **thermophiles** if you have already cloned the gene from an organism such as *E. coli* which is easier to grow and manipulate but not so biotechnologically useful.

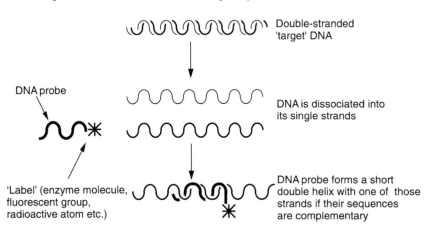

Double-stranded 'target' DNA

DNA probe

DNA is dissociated into its single strands

'Label' (enzyme molecule, fluorescent group, radioactive atom etc.)

DNA probe forms a short double helix with one of those strands if their sequences are complementary

Traditionally, DNA probes have been made by cloning a gene and using its DNA as the probe. In the last few years oligonucleotides made on a DNA synthesizer have gained favour as probes. They react faster, so reducing assay time, they can be made more specific, so distinguishing between genes which differ by only one base, and they can be made in comparatively large amounts very cheaply. Indeed, the primers necessary for such techniques as PCR could be considered as a form of probe.

See **Hybridization, Oligonucleotide**.

DNA sequencing

Determining the sequence of the bases in DNA (DNA sequencing) is one of the mainstays of gene cloning technology. There are two common methods for doing this:

● Maxam and Gilbert technique (chemical degradation). This uses chemicals to break the DNA into fragments.

● Sanger technique (di-deoxy method, chain termination method). This uses enzymes to make a new DNA chain on the target you want to sequence, using the 'di-deoxy' reagents to stop the chain randomly as it grows.

In both cases the results of a series of reactions are analysed by polyacrylamide electrophoresis, to give information which can be read directly to give the sequence of the original DNA.

An associated technique is m13 cloning. m13 is a small virus which infects *E. coli*, and which is particularly convenient for making short sections of DNA to sequence. One favoured way for sequencing large pieces of DNA is to chop up the DNA chain into random pieces, clone each piece by splicing it into the m13 virus, and then sequence viruses at random until you have covered all of the original DNA sequence. This is called 'shotgun' cloning or sequencing.

The human genome project, the project to sequence all three billion bases of DNA in man, has led to a lot of interest in building robots to sequence

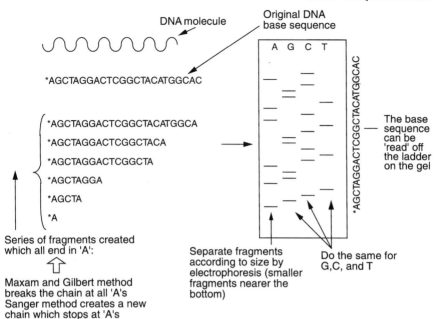

DNA. So far automated machinery only handles discrete parts of the sequencing process, and many of the most advanced laboratories continue to do sequencing 'by hand', claiming that the results are much more reliable.

See also **Genome project.**

Downstream processing

This is a general term for all the things which happen in a biotechnological process after the biology, be it fermentation of a microorganism or growth of a plant. It is particularly relevant to fermentation processes, which produce a large amount of a dilute mixture of substrates, products, and microorganisms. These must be separated, the product concentrated and purified, and converted into a product which is useful.

DOWNSTREAM PROCESSING

There are three general steps to downstream processing:

- Separation
- Concentration
- Purification

(*see* **Separation, Concentration, Purification**). The first step separates the crude product from the microbial mass and other solid lumps, the second removes most of the water (and hence is often called dewatering), the third takes the concentrated product and purifies it. The orders can be different, but generally fall into this scheme.

Separating the microbial mass is necessary whether the product is inside the microorganism or outside it — the difference is that in the first case you keep the mass, in the second you throw it away. This can be done by centrifugation (expensive but guaranteed efficient), filtration methods, especially cross-flow filtration (*see* **Cross-flow filtration**), or by flocculation (adding something to the microbes so that they clump together and settle out on their own). If the product is inside the organism then separation also concentrates the product: however you then have to break open the organisms to get it.

Some of the same methods can also be used for concentration. Simply drying large volumes of liquid is usually too expensive, so again ultrafiltration or reverse osmosis (both membrane methods which keep the product on one side of the membrane while most of the water goes through it to the other) are popular.

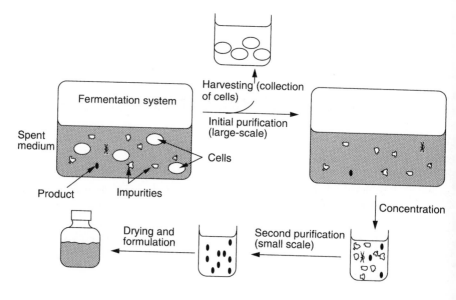

Product concentration. The result of the above steps is usually a rather dilute solution of the product, which must be concentrated. This can be achieved by reverse osmosis, adsorption methods, extraction with another liquid.

Purification. Most biotechnology products are produced as mixtures by cells, but are required in pure form. Purification methods include chromatographic, affinity methods, various specific precipitation methods. If the product is produced by genetic engineering, then it may be engineered to have a molecular 'hook' which makes it easier to isolate.

See also **Cell disruption.**

Drug delivery

This is the method by which a drug is delivered to its site of action. For traditional drugs this is a different name for formulation, i.e. in what form the drug is to be given to the patient (tablet, capsule, syrup, etc.). A drug can also be made as a prodrug, a compound that is not itself a drug but which the body metabolizes into the drug. If the metabolism only occurs in one tissue or cell type, then the drug is only released there. Although a skilled area of pharmacology, this has little biotechnological content. However two aspects of biotechnology have needed drug delivery technology.

Firstly, biotechnology has allowed the development of a range of new drug delivery systems, such as **liposomes** and other encapsulation technologies, and antibody-based drug targeting mechanisms (such as immunotoxins) that target a drug to a particular cell or tissue.

Secondly, biotechnology has also created the need for new drug delivery systems to get biotechnology-derived drugs to their site of action. This is particularly acute in the case of biopharmaceuticals, protein drugs which cannot be taken by mouth as the stomach acid and gut enzymes would destroy them. Even if they could survive this digestive gauntlet, they would rarely get into the bloodstream, because the protein molecules are too big to diffuse through the intestinal walls.

DRUG DELIVERY

The current solution is parenteral delivery (i.e. injection): this works well, and has been the method for giving patients insulin (a protein drug) for decades. However it is invasive, expensive, and carriers a continual risk of infection or tissue damage. Thus several biotechnology companies have set out to find a better way of getting proteins into the blood. There are several approaches.

Transdermal delivery. This uses a method of getting the protein across the skin without knocking visible holes in it. Methods used include iontophoresis (using electric fields to push the drug across the skin) and high pressure jets of fluid. However the skin has evolved to resist exactly this sort of attack, so these methods are not very effective for proteins.

Oral delivery. Taking the drug by mouth, but with some other materials which help it survive the gut. These can include protease inhibitors (to block the digestive enzymes), or carrier materials which protect the proteins but which dissolve at the right time to make them available for absorption. Other tricks include linking the protein to something, like vitamin B_{12} which is actively taken up from the gut, so that the protein is absorbed along with it.

Nasal/pulmonary delivery. The cells lining the lungs and part of the nose (their epithelial cells) are very poor barriers to proteins compared to skin and guts, and so are potential 'weak points' for drug delivery. The nose is particularly attractive, because it has a large internal surface with lots of blood vessels, and is easily accessible.

Protein remodelling. This approach attempts to remodel the protein chemically to protect it against the difficulties of entering the body. This can be done by encapsulating it (as above), by embedding it in a variety of carrier materials such as dextrans, albumin, xanthan gums, or synthetic polymers such as polyethylene glycol (PEG), or chemically modifying it with these or other materials.

The blood–brain barrier. Many chemicals in the blood do not affect brain and spinal cord nerve cells. The nerve cells get their nutrients from surrounding cells and from the cerebrospinal fluid (CSF), which is not part of the circulatory system of the rest of the body. The cells form a barrier to penetration of drugs in the blood into the nerve cells of the brain. This can be a problem, as taking drugs by mouth or even injecting them is much safer and easier than injecting them into the CSF. A substantial part of the effort in drug delivery is remodelling the drugs so that they can penetrate this blood–brain barrier.

So far, protein drug delivery systems have been much hyped, but not very effective. It is not clear whether they ever will be, or whether re-designing biopharmaceutical drugs to be chemically more robust and better suited to getting into the body will be successful before the drug delivery systems are launched into general use.

See also **Immunotoxins**.

Drug development pathway

A substantial amount of biotechnology is concerned with developing new drugs, often biopharmaceuticals. As a consequence, some of the jargon of drug development and licensing creeps into biotechnology discussions. This entry summarizes the key points in the pathway that a new drug candidate has to go through.

Preclinical research. This means all the research that goes on before you try the compound out on people, but is often taken to mean animal studies of the drug. Studies using biochemical methods, receptor screening, or cell culture assays are usually considered just 'research', as most of the drug candidates they throw up will not make it even as far as clinical trials.

Phase I trials. These are the first trials in which a drug candidate is given to people. The only permission needed to do Phase I trials is that of the local hospital ethical board or committee (which, however, has to be convinced that there is some point in doing the trial). The people are normal, healthy volunteers (often medical students), and the purpose of the trial is to establish the pharmacokinetics of the drug, and to find the minimum dose which will have some effect: thus the trial starts out with a very small dose and works up. Usually only a small number of people — between 10 and 20 — are involved.

After Phase I the developer applies for an Investigational New Drug application (IND, in the US), or the equivalent in other countries (e.g. the Clinical Trial Exemption certificate — CTX — in the UK). This is the regulatory hurdle necessary to go on to Phase II trials, and at this point the developer must show that they have complied with a wide range of Good Laboratory Practice (GLP) rules in their pre-clinical and Phase I trials . For medical devices such as prostheses (whose development path is essentially the same), the IND is replaced by the 510(k) application in the US.

Phase II trials. This is the first time the drug is tried on ill people. This trial is usually done at one hospital centre on a small number of patients, and looks for any evidence that the drug actually has a medical effect on the disease that it is meant to treat. The drug is said to be being developed for one 'indication', i.e. one collection of symptoms or one disease type. The object of this and subsequent trials is to show that the drug has an effect *for this indication.* (Note that up to this point tests can be for any disease.) Again, the numbers of patients are quite small.

Phase III trials. This is where very big money is spent on drug development. The object of this phase is to see whether the drug is worth launching, because it is better than existing therapies, does not have severe side-effects and so on. This requires hundreds or thousands of patients (all of whom have to be followed in detail), usually at at least six hospital centres. The trial is done double blind, so that neither the people giving the drugs nor the people analysing the results know who has received the drug and who has received a placebo (the treatment, be it a pill or injection, that does not contain the new drug) until the study is over. It is also usually a crossover trial, i.e. one in which, half-way through, those who had been receiving placebo are given drug and those who had been receiving drug are given placebo. (This helps to avoid problems due to the differences that people show in their response to drugs.)

At the end of Phase III the drug is submitted for a New Drug Application (NDA, in the US) or a Product Licence Application (PLA in Europe). (For a medical device, the equivalent is the Pre-marketing Approval — PMA.) If this is approved, then the drug can be sold.

Phase IV trials. However, selling it does not mean that development is over. Phase IV trials — post-marketing surveillance — then takes over to look for rare adverse reactions, to look for opportunities to decrease the dose (because the initial estimates, derived from Phase III studies, are often rather high), and to extend the range of indications for which the drug may be used. Extension of indications can come about because of *off label* use, that is use of the drug by physicians for indications other than those for which the drug is licensed. There is nothing to stop people doing this, providing that they are very careful to emphasize to their patients that they are effectively doing an experiment on them. Successful experiments lead to new ideas for the use of the drug, and hence new clinical trials to see if the new indication is a suitable one for this drug.

See **GLP/GMP**.

Electrochemical sensors

These are types of **biosensor** in which a biological process is harnessed to an electrical sensor system, making a sensor. The most commonly discussed type of electrochemical sensor is the enzyme electrode (*see* **Enzyme electrode**). Other types couple a biological event to an electrical one via a range of mechanisms. Among the more common are the following.

Oxygen electrode-based sensors. These are sensors in which an 'oxygen electrode' (Clark electrode), a standard electrochemical cell which measures the amount of oxygen in a solution, is coated with a biological material which generates or (more usually) absorbs oxygen. When the biological coating is active, the amount of oxygen next to the electrode falls and the signal from the electrode changes. Typical coatings might be an oxidase enzyme (which consumes molecular oxygen to oxidize a particular substrate) or a whole cell (which consumes oxygen when presented with a range of substrates). This latter type of biosensor — a microbial- or cell-based biosensor — can be used to detect poisons, as poisons damage the cells and hence reduce the rate at which they consume oxygen.

pH Electrode-based sensors. Again, a standard electrochemical pH electrode is coated with a biological material. Many biological processes raise or lower pH, and so can be detected by a pH electrode. Examples could include hydrolysis of an ester to acid and alcohol, or, again, the metabolism of neutral pH substrates by a bacterium. Indeed, in one study which was meant to measure the pH inside a volunteer's mouth by placing a very small pH electrode there, what the electrode ended up detecting was sugar. Bacteria grew over the electrode and, every time the subject ate sugary food, the bacteria metabolized some of it to lactic and acetic acid and the pH next to the electrode fell from 7 to 4.5.

Electroporation

This is manipulating cells by exposing them to a strong electric field. Initial studies (as one would expect) showed that while one could manipulate cells with strong electric forces, the cells never survived the treatment. However by altering the conditions suitably, electroporation can be used to transform cells with DNA and to fuse cells.

Transformation of cells — getting DNA into them — can be achieved simply by exposing the cells to a suitable electric field while they are in a solution of the DNA. The electric field seems to modify the lipid membrane that surrounds the cells, and greatly increase the rate at which pinocytosis, a normal mechanism by which cells take up chemicals from solution, takes DNA into the cell. It is not widely used for animal or bacterial cells, where other methods have been developed which are quite reliable. However, electroporation is discussed fairly extensively when talking about getting DNA into plant protoplasts, and to a lesser extent into fungal cells. Some workers claim that electroporation or electrophoresis can even get DNA into intact plant cells (i.e. cells with their cell wall still there): the evidence for this is generally considered to be poor.

Fusing cells was the first application of electroporation. Protoplasts of plant cells or whole animal cells, can be made to fuse by putting them next to each other and exposing them to a strong electric field. There seems to be no limit to the types of cell which may be fused together using this technology. Early studies uniformly resulted in dead cells: however techniques have improved now so that cells may be fused to produce viable offspring using electroporation. Applications in plant genetics include making hybrid plants and polyploid plants. These latter are plants which contain more than the usual number of chromosomes (usually two or three times as many as normal).

Embryo technology

Embryo technology is a generic name for any manipulation of mammalian embryos, and is relevant to biotechnology in two areas. Firstly, biotechnological methods and materials make some of the embryo technology possible. Secondly, biotechnological techniques such as transgenic technology rely on embryo technology to provide them with tools of the trade. Embryo technology encompasses:

- Cloning. This could be done in two ways, in principle: by embryo splitting (see below) or by nuclear transplant. The latter method takes the cell nucleus from an adult cell and places it in a fertilized egg from which the nucleus has been removed. The egg then goes on to develop using the genetic material from the adult cell. As there are billions of cells in any adult mammal, this provides a route for making a billion-strong clone of one person. Or it would do if it worked, but it only seems to be at all reliable on frogs, and even then the most skilled scientists in the field can only get it to work sometimes.

- Embryo splitting. This is taking an embryo when it consists of only a few cells and splitting it into smaller bundles of cells. As many as eight new embryos can be made this way — if you split a mammalian embryo any more than that then the resulting clusters of cells do not develop into fetuses.

- *In vitro* fertilization. A widely used technique for animals and people, this means fertilizing the egg with sperm outside the body. Usually the fertilized egg is cultured outside the body for a few days before re-implantation into a female to make sure that fertilization has occured. IVF has been the subject of intense and emotional public debate since it became possible to do it on humans in the early 1980s. A related technique is GIFT, which injects sperm directly into the Fallopian tubes, and so is a half-way house to the fully external fertilization of IVF.

- Artificial insemination (AI). This is simply fertilizing a female with sperm from a male without copulation. It is widely practised on people, farm animals, fish, oysters, and many plant species (although it is not usually called AI in the last case).

EMBRYO TECHNOLOGY

- Gamete and embryo storage. This is the storage of eggs, sperm, or fertilized embryos outside 'their' original source (an animal or person). Almost invariably this means freezing them at liquid nitrogen temperatures. There is also intense ethical debate about this.

Two of the other key points for debate about embryo technology are:

DNA-based genetic diagnostics. Because DNA probes can detect 'defective' genes whether they are actually doing anything or not, they can and have been used to detect whether a fertilized egg, an embryo, or a fetus is carrying an undesired gene. If it is, then it can be aborted before the gene has a chance to express itself. This technology is often tied up with the debate about the ethical acceptability of abortion. Almost all *in utero* diagnostics (i.e. diagnostics done on a growing fetus in its mother's womb) are done to allow the mother to decide whether she wishes to terminate the pregnancy. There are no cures for the diseases that the DNA technologies detect, and no treatments for them that cannot wait until the baby is born. Thus the only reason for doing the DNA test is to decide whether to have an abortion, and so antiabortion campaigners see *in utero* DNA testing as being part of the abortion technology.

When is a fetus ...? The ruling in the UK, following the influential and generally accepted Warnock Report, is that an embryo is not recognizably human until 14 days — before this it is classed as a 'pre-embryo'. After 14 days it becomes an embryo, and starts to acquire some rights as a human being. Sometime between then and around 15 weeks the embryo is renamed as fetus. The fetus is not usually considered capable of independent life until 24 weeks gestation (and even then only with heroic medical intervention, and a high risk of 'congenital' malformation). By 35 weeks gestation the fetus is generally capable of independent life if cared for in a specialist premature baby unit ('Special Care Baby Unit', SCBU, pronounced Skiboo). Clearly somewhere between fertilization and 35 weeks gestation the developing pre-embryo/embryo/fetus has become human. There is wide debate as to when this occurs, and whether it occurs at one time or as a continuing process.

See also **Yuk factor.**

Embryogenesis
(in plant cell culture)

Embryogenesis is encouraging plant tissues to form new plants *in vitro*. The original experiments in the late 1950s showed that small pieces of carrot tissue could be grown back into whole carrot plants by culturing them in sterile conditions with the right chemicals. The new plants are usually very similar to the embryo plants which first emerge from the seeds, so this represents the cells returning to the 'genetic program' at the start of the plant's life cycle. Although this usually only happens with the seed cells (germ cells), the embryogenesis referred to here is somatic cell embryogenesis, i.e. making embryos from cells outside the usual reproductive apparatus. Quite a number of plants occasionally generate embryos without generating seeds, so doing it in cell culture is exploiting a mechanism present in most, maybe all, plants.

The generation of embryos takes two stages: initiation and maturation. The former needs a high level of the group of plant hormones called auxins: the latter needs a lower level. Other chemicals have to be at suitable levels, too. Thus the procedure is usually to take a piece of plant tissue and put it on a high-auxin medium, where the cells grow into a mass of 'callus'. This is then transferred to a maturation medium, where the callus starts to develop initial organs, ultimately growing a root and a shoot.

In plant culture circles, embryogenesis is used to describe the generation of new plants from bits of old plants. If you generate a plant from a single cell, it is organogenesis, although the techniques have many similarities. Embryogenesis is critical for plant cloning and micropropagation technologies.

Encapsulation

This is any method which gets something, usually an enzyme or bacterium, into a small package or capsule while it is still working (or alive). The capsule can be any size, but usually is no bigger than a few millimetres across. If it is too small to see fairly easily with the naked eye, the process is called microencapsulation.

Encapsulation is one method for immobilizing cells for use in a bioreactor. Encapsulating agents can be anything which will form a shell around something, but usually are polysaccharides such as alginate or agar, because they are inert, allow nutrients and oxygen to diffuse into and out of the sphere readily, and are easy to convert from gel (solid) to sol (liquid) or solution form by altering the temperature or the concentration of ions such as calcium. Proteins such as collagen (gelatin) are also used.

Enzymes may also be encapsulated, although they are more usually immobilised on the surface of polymer particles

Drugs are often encapsulated to help their survival or release in the patient. A variety of sustained-release cold cure medicines which come as little particles inside a capsule are actually encapsulated drugs: each particle contains a shell of a slowly dissolving material around a core of solid drug powder. Only after the shell has dissolved in the intestines can the drug get out to the body. By having a mixture of shell thicknesses the pharmacologist arranges for the drug to be released over a period of time. Similar approaches have been tried for biotechnological drugs, although not always very successfully. Encapsulation of drugs is also a method of protecting them from, say, acid in the stomach so that they can be taken by mouth instead of having to be injected. This has been something of a 'holy grail' of drug delivery technology for biopharmaceuticals, but has not been successful to date.

See also **Enzyme immobilization, Liposome**.

Environmental biotechnology

Environmental biotechnology is a general term covering any biotechnological product or process which can be considered to be helpful to 'the environment'. Usually, this means control, reduction or disposal of waste, removal of chemical pollutants, or reduction in power use, especially in industry. Because of the high political profile of 'the environment', a number of diverse biotechnology activities have been included under 'environmental biotechnology'.

Biotechnology is well-placed to address some ecological and environmental concerns. As opposed to traditional heavy industry, biotechnology is likely to use potentially renewable resources, inherently low power processes, materials which are unlikely to be dangerous, and to produce products which can be justifiably labelled as 'natural'.

The most commonly discussed topics in environmental biotechnology are:

- Bioremediation: cleaning up contaminated soil using biological processes (*see* **Bioremediation**).

- Soil amelioration: improving soil quality through manipulation of its microflora (*see* **Soil amelioration**).

- Developing biodegradable replacements for plastics, and particularly developing biotechnological ways of making them (*see* **Biodegradable materials**).

- Waste disposal: developing bacterial methods for disposing of waste, or at least of disposing of the biodegradable part of it, more rapidly.

- Creating alternative energy sources: specifically biofuels, biogas, and solar energy methods (*see* **Biofuels, Biogas, Solar Energy**).

Enzyme Commission (EC) number

All enzymes are given a systematic name and number which identifies them in technical literature. (They may also have a 'common' name, like trypsin or rennin.) These names are given by the Enzyme Commission. The name and number are systematic descriptions of what the enzyme does. The number is a four-digit number. The first digit classifies the enzyme into one of six broad groups:

Number	Class
1	Oxidoreductases (transfer of H atoms or electrons)
2	Transferases (transfer of small groups between molecules
3	Hydrolases
4	Lyases (addition to double bonds)
5	Isomerases
6	Ligases (formation of bonds between C and another atom, using ATP as an energy source)

Each of the groups is subdivided into subgroups, each subgroup into subsubgroups, and the last number is specific for the enzyme. The systematic name describes the reaction catalysed. Thus creatine kinase is EC 2.7.3.2 (2 because it transfers a group from ATP to creatine, 2.7 because the group is phosphate, 2.7.3 means the subgroup that transfers the phosphate to a nitrogen atom). Note that the dots are essential, as some classes of enzyme have more than ten members. The systematic name is ATP:Creatine phosphotransferase — the enzyme that transfers a phosphate group from ATP to Creatine.

Enzyme electrode

A type of **biosensor**, in which an enzyme is immobilized onto the surface of an electrode. When the enzyme catalyses its reaction, electrons are transferred from the reactant to the electrode, and so a current is generated. (This is distinct from other types of electrochemical biosensors, where the enzyme generates a distinct chemical product, for example an acid, which is then detected by a separate electrode system.)

There are two types of enzyme electrode:

- Ampometric. Here the electrode is kept as near zero voltage as is practical. When the enzyme catalyses its reaction, electrons flow into the electrode, and so a current flows.

- Potentiometric. Here the electrode is held at a voltage which counteracts the voltage created by the enzyme's tendency to 'push' electrons into it. This may be done by actively adjusting the voltage or by not connecting the electrode to anything (as is the case in ISFET devices). The device's output is the voltage necessary to prevent any current flow through the electrode.

Usually enzymes transfer their electrons inefficiently to the electrode, so a mediator compound is coated onto the electrode to help the transfer. The favoured mediators are ferrocenes, because they can easily carry a single electron at the electrode potential suitable for enzyme oxidations and reductions. A range of other organic chemicals have been considered, and the 'organic metals', i.e. organic compounds which conduct electricity, hold promise as electrode materials. Ionomers are also used. These are polymers which are not charged (and so stick to the electrode) but which have a charged group as a side chain.

The enzyme has to be immobilized on the electrode in some way. Common methods include:

- Physical adsorption. The enzyme is encouraged to stick to the enzyme surface. Many proteins will stick quite avidly to some surfaces, held there by small patches of electrostatic charge or because they sit in a hydrophobic 'pocket'. This approach is simple, but the enzyme can simply leach off again unless it is held very tightly (which they usually are not).

- Chemical crosslinking. The enzyme is chemically linked onto the electrode surface. Rarely do the chemistries of the enzyme and the electrode match to allow this route.

- Immobilization in a gel. The enzyme is mixed with a polymer such as agarose or polyacrylamide, and then chemically crosslinked to the gel to form a solid capsule around the electrode.

- Capture behind a membrane. Here the electrode is inside a small sack which is permeable to the analyte but not to the enzyme. The enzyme is inside the sack.

A vast number of enzyme electrode systems have been developed in laboratories, and the early 1980s saw a boom in interest in their application. However they nearly all proved to be hopelessly impractical for commercialization. The major exception was the glucose biosensor for diabetic monitoring: a few other medical sensors are now being commercialized.

Enzyme mechanisms

As the use of enzymes is one of the most commercially important areas of biotechnology, understanding how they work is an important part of the research underpinning the technology. Indeed, one of the reasons why enzymes are so widely used is that their mechanism of action has been studied for almost a century, and the science of enzymology is correspondingly mature (as opposed to, say, the relatively new science of molecular genetics).

Specific aspects of how enzymes work and how they may be improved for a particular application are dealt with elsewhere. The fundamental research involved is beyond the subject (and scope) of this book. However, there are several lines of research which use relatively new technologies in enzymology:

- Chemical modification. Changing an amino acid in the protein into another by chemically reacting it. This usually results in a change in the enzyme's activity, and if it does the change is almost always 'for the worse', that is it reduces the enzyme's catalytic effect, specificity, or both. Sometimes the change can result in a commercially more useful enzyme, in which case the modified protein is used commercially. However the protein is changed, the result is usually interesting to the enzymologist.

• Site-directed mutagenesis. Changing an amino acid into another amino acid by genetic modification. This is more flexible than chemical changes, because an amino acid, identified maybe from protein sequencing work or X-ray crystallography, can be changed specifically to another, closely related (or totally unrelated) amino acid (*see* **Site-directed mutagenesis**).

Enzyme production by fermentation

Industrial enzymes may be made by extraction from a naturally occuring source, often part of a large animal or plant, or by production from a microorganism in fermentation. The former requires less equipment, but is more prone to seasonal variation and the vagaries of climate, international trade, and (in extreme cases) war and revolution interrupting supply. Fermentation has the potential to provide a more uniform and reliable source of material.

The enzymes which account for the majority of production are essentially commodity products. Thus a substantial part of the cost of their production is the raw materials and power necessary to produce them. (This differs from enzymes used for research, such as restriction enzymes, which are produced in comparatively tiny amounts and whose production cost is dominated by the skilled labour necessary to make them, *see* **Recombinant DNA: bits and kits.**) Thus a successful fermentation process must use low cost feed materials, an organism which does not require excessive heating or cooling, and one which makes a lot of the enzyme.

Typical feedstocks are hydrolysed starch, molasses, whey, and cereals for carbon, soy flour, fish meal, blood, and cotton-seed meal for nitrogen.

ENZYME PRODUCTION BY FERMENTATION

For high-value enzymes (for example for use as drugs), some of these are inappropriate as they contain insoluble dirt which will have to be rigourously removed from the final product. Fermentation conditions which must be monitored to optimize enzyme production include pH, oxygen, CO_2, aeration, temperature, agitation, and, as some enzymes are denatured on surfaces or may concentrate at them, foaming. In addition, many enzymes' production by bacteria is induced and repressed by specific chemicals. Inducers must be present and repressors removed from the fermentation if the yield is to be satisfactory.

Many industrial enzymes are sold as fairly crude preparations, with a mixture of proteins in them. These have been prepared by separating the cells from the fermentation broth, and then partially purifying the protein from the liquor by precipitation, ultrafiltration, or a similar technique.

See also **Induction**.

Enzyme stabilization using antibodies

This is a method for stabilizing proteins, usually enzymes, by binding antibodies to them. Some enzymes may be stabilized 200-fold by complexing them with an antibody, i.e. the 'half-life' of their enzymatic activity can be increased (from 5 min to 16 h in the case of α-amylase, for example). The antibodies have to be selected so that they do not block the active site of the enzyme, as otherwise the protein is stabilized but becomes inactive as a catalyst: thus monoclonal antibodies, which bind to defined bits of the protein's surface, are usually used.

The process works because the antibodies bind to the active structure of the enzyme. If the enzyme now 'tries' to unfold into an inactive structure, it must not only overcome its own binding energy but also throw off all the bound antibodies. This requires more energy, and so is a correspondingly slower process.

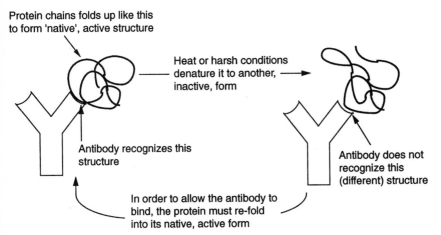

Protein chains folds up like this to form 'native', active structure

Heat or harsh conditions denature it to another, inactive, form

Antibody recognizes this structure

Antibody does not recognize this (different) structure

In order to allow the antibody to bind, the protein must re-fold into its native, active form

Antibody stabilization is used routinely to stabilize enzymes used in medical diagnostic assays. The antibodies are too expensive for this process to be routine for enyzmes used in large-scale processes.

Enzymes

The core of traditional biotechnology, and a key feature of the new bio-technology of gene cloning, is the use of enzymes. For practical purposes these can be considered to be catalytic proteins, although recent work has shown that RNA can act as an enzyme.

Enzymes are prepared from a huge variety of organisms, from viruses to whales. In general, they may be extracted from some organism which already produces the enzyme, extracted from a microorganism which is cultured under conditions in which it produces the enzyme, or made from an organism which has been genetically engineered to produce the enzyme.

Enzymes are so widespread in biotechnology that they crop up in many entries in this book. Specific classes of enzymes covered are **glycosidases, glucose isomerase and invertase, proteases**, and **lipases**. Enzymes also crop up in, **biotransformation, protein engineering, enzyme production by fermentation, enzyme mechanisms**, and **expression compartment**, as well as in many other entries.

The value of enzymes to the biotechnology industry can be estimated from the following table.

Industrial enzyme	Market value (US$millions)
Pharmaceutical proteins	100[*]
Detergents (proteases and lipases)	70[+]
Dairy industry (mostly rennin)	50
Research (a wide variety of enzymes)	42
Starch processing	31[‡]
Diagnostic (a wide variety of enzymes)	16
Textile processing	12[§]
Drinks industry	11
Baking (*see* '**Glycosidase**')	4.5[¶]
Biotransformation	4.5
Others	5
Total	400 (for 1990)

[*]This includes enzymes like TPA (*see* **Blood products**). [+]Protease detergents are the traditional enzymes, although the fat-dissolving lipases are beginning to be used in retail as well as industrial detergents now. [‡]*See* **Glucose isomerase and invertase, Polysaccharide processing, and Glycosidases**. [§]Proteases and cellulases: cellulases and amylases used for whitening and softening cotton (for, for example, producing 'stone-washed' jeans). [¶]Variety of glycosidases for improving dough quality.

Expression compartment (inclusion bodies)

Getting a protein made in a recombinant cell is relatively straightforward, as a wide range of expression vectors exist that can be used to clone the relevant gene. However often the protein is produced in a form which is

EXPRESSION COMPARTMENT (INCLUSION BODIES)

not useful to the genetic engineer. This is often a feature of where in the cell the protein is made.

Inclusion bodies. These are condensed particles of protein which are formed inside bacteria and (to a lesser extent) eukaryotic cells when the cells are forced to make large amounts of protein. The proteins are often crosslinked and/or denatured, making it useless for its intended purpose. Inclusion bodies were the bane of early recombinant DNA production methods, but the skills required for manipulating bacterial physiology (i.e. how they grow) to avoid inclusion bodies are now much better developed.

Getting your protein as an inclusion body is not a catastrophe. Such proteins can be refolded by dissolving them in a detergent or 'chaotropic agent' solution, and then gradually removing the detergent by dialysis. Using the right buffers, this allows the protein to refold into its proper form. However this is a bit of a 'black art', and does not always work.

Cytoplasmic expression. If you do not tell a protein where to go in a cell, it stays in the cytoplasm (the space inside the cell wall). Most proteins are expressed in the cytoplasm — however this is where inclusion bodies form, and also where an efficient mechanism for breaking down aberrant proteins exists. As far as the cell is concerned your genetically engineered protein is aberrant, so it can be broken down very rapidly in the cytoplasm. (This is especially true for very small proteins or peptides — large ones tend to form inclusion bodies instead.)

Periplasmic space. This is the space between the cell membrane and the outer cell wall in bacteria. Many proteins which are secreted (see 'secretion' below) end up here. This has the advantage that it gets them out of the cytoplasm, but does not release them free into the medium (so that they can be harvested simply by collecting the cells). However the periplasmic space has its own set of digestive enzymes which can break down proteins, aimed at completely different types of protein molecule to the cytoplasmic ones.

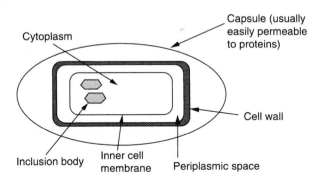

EXPRESSION COMPARTMENT (INCLUSION BODIES)

Secretion. Some proteins can be exported from the cell entirely, to remain free in solution in the culture medium. This makes it much harder to collect them, but if they can be exported successfully they rarely precipitate as inclusion body-like lumps. There may, however, still be proteases in the solution which can break them down.

Expression systems

Usually a cloned gene will be inert: it will not perform its usual function in the host cell, as it is outside its usual genetic context. Expression systems are combinations of host and vector which provide a 'genetic context' which makes the gene function in the host cell — usually, this means that it makes a protein at high levels.

Because making lots of foreign proteins is often lethal to the host cell, there are several variations on the theme of the expression vector which allow the level of the protein made from a cloned gene to be increased:

• Inducible systems. Here the expression of the cloned gene is 'turned on' by an inducer, so that the cells can be grown up in bulk and then induced to make the protein.

• Amplification systems. Also called high copy-number vectors. Usually the plasmids and viruses from which vectors are made are present in only a few copies per cell.

High copy number vectors are present in hundreds of copies. More genes lead to more protein produced. The increase in the number of genes can be made conditional on, say, a rise in temperature, so that the host cells are grown at one temperature and then fill up with DNA and the target protein at another temperature.

• Runaway replication plasmids. This is a logical extension of the amplification system. When the temperature is increased, the normal system which controls how much plasmid DNA is present is destroyed and the bacterium continues to make plasmid DNA until it runs out of precursors to make it from. This results in a cell full of plasmid, and hence, in principle, with the product of its genes.

- Secretion vectors. These are vectors which allow the protein product of the cloned gene to be secreted from the cell. This can be very helpful for purification, as all the other proteins in the host cell are removed with the cell itself, but does not always work because the target protein is broken down in solution, is not stable, or is incapable of being secreted effectively.

Even with a host cell and vector which are consistent with the gene you want to express, actually obtaining high levels of expression can be difficult. Obtaining a fraction of a per cent of the cellular protein as the product you want is a research target and easy to achieve. However, obtaining the 10 per cent or more of target protein, which is necessary for economic production of any but the most valuable proteins, can be thwarted by unforeseen effects of such high levels of protein on the cell itself, requiring the biotechnologist to switch to another expression system, often switching from bacteria to yeast or to mammalian cells.

Another common problem with expression systems is that of the formation of inclusion bodies, where the protein is accumulated as an insoluble, inactive mass inside the cell rather than in its native, active form.

Thus obtaining the best performance from an expression system requires considerable knowledge of how the host cell's internal machinery (its physiology) works.

A novel approach to expressing foreign proteins is to use transgenic animals. Here, rather than a bacterium or a yeast, a mammal is the carrier for the foreign gene, which is spliced onto the front of the gene for lactalbumin, a major component of milk. The animal expresses the gene construct in the mammary glands, and the recombinant protein is secreted in relatively pure from into the milk. GenPharm is a transgenic company specialising in producing pharmaceutical proteins in this way. Production of pharmaceutical proteins from transgenic animals' milk is sometimes called 'pharming'.

See also **Expression compartment, Induction, Secretion, Transgenic animals: Applications**.

Fermentation processes

Strictly speaking, fermentation is microorganism metabolism under anaerobic conditions on a carbon substrate. However it has been extended to mean growing microbes in liquid under any conditions. Growing cells in small amounts on a petri dish or in small-scale mammalian cell culture is called incubation, and takes place (unsurprisingly) in an incubator.

There are three general ways in which fermentations are done, each with a variety of associated terms. In all cases there are some common terms in bacterial growth, such as the bacterial doubling time (the time needed to double the number of bacteria there, *see* **Cell growth**).

Common terms. For all bioreactor processes, the first thing that happens is that the bioreactor is sterilized. This can be done with steam, chemicals, washing or some combination of these. The fermentation is then started with an inoculum, a small, actively growing sample of the organism to be cultured. Fermentation then proceeds according to one of the schemes below.

- Batch fermentation. Here the reactor is filled with a sterile nutrient substrate and inoculated with the microorganism. The culture is allowed to grow until no more of the product is being made, when the reactor is 'harvested' and cleaned out for another run. The culture goes through lag phase (when the organisms adapt to their surroundings), exponential growth, when they grow in numbers, stationary phase, when they stop growing, and death phase. Depending on what the product is, the 'useful' part of the growth cycle can be any one of these four stages, although it is usually the growth or stationary phases.

- Fed batch fermentation. Here the batch culture is fed a batch of nutrients before it gets to the stationary phase, so that it never runs out of nutrients. At the same time some of the fermentation is removed and taken off for processing.

- Continuous culture. This is the logical extension of fed batch fermentation. The fermenter is fed continuously with nutrient and the culture medium removed continuously. This has some advantages over fed batch systems in that the culture conditions are always the same, but also is harder to control. This is essentially a large-scale chemostat.

Fermentations may also be classified according to when the product is made:

- Type I fermentation — product is made from primary metabolism.

- Type II fermentation — product is made from secondary metabolism at the same time that primary metabolism is going on (i.e. when the cells are growing).

- Type III — product is made by secondary metabolism at a different time from the primary metabolism (i.e. during stationary or death phases of the culture).

Lastly , fermentations may be classified according to how the culture is kept 'clean'.

- Aseptic/sterile fermentation. All other organisms are excluded by the biotechnologist. This is by far the most common case.

- Consortium fermentations. Here a group of organisms are growing together, rather than just one organism. For this to work, the organisms must be dependent on one another, as otherwise one will outgrow the others and dominate the culture.

- Protected fermentations. Here the culture is not aseptic, but is performed under conditions under which only one type of organism will grow. Thus fermentations at extremely high temperatures, extremes of pH, or with very hard-to-metabolize substrates will tend only to be able to support the organism that the biotechnologist is after, thus removing the problem of having to keep contaminants out.

Fermentation substrates

Many materials are used as food for growing microorganisms. These are referred to as substrates. The substrate and the trace materials needed, together with chemicals added to make the fermentation easier (such as anti-foam agents to stop froth forming) make the culture medium.

FERMENTATION SUBSTRATES

The substrates can be divided into those providing the different essentials for life: a source of carbon, nitrogen, and (in the case of aerobic fermentation) oxygen. Usually carbon substrates cost the most, because you need most of them. Among common carbon substrates are:

- Molasses. A side product of sugar refining that contains most of the material from sugar beet or sugar cane which is not sugar, molasses is one of the cheapest substrates available.

- Malt extract. Made from malted barley by soaking it in water.

- Starch and dextrins. Polysaccharides often made from cheap crops like potatoes.

- Cellulose. The world produces about 100 billion tonnes of cellulose a year, so it is a potential raw material for large-scale fermentation. But only a few organisms can degrade it.

- Whey. A side-product of dairy processes, it is cheap but expensive to store or transport.

- Methanol. A very cheap chemical from the oil industry, but it contains no nitrogen. Only a restricted range of organisms can grow on methanol. Similarly ethanol ('alcohol') can be used, but more usually ethanol is the product of a fermentation.

- Oil, gas. Some organisms can use natural gas or some components of crude or refined oil as carbon substrates. However their commercial use depends critically on the price of oil.

Nitrogen substrates include:

- Ammonia. A very smelly gas produced as a bulk commodity for the chemical industry. Most organisms can use ammonia. Sometimes it is converted into ammonium salts or into urea for ease of handling.

- Corn steep liquor. The residue left behind when making starch from maize.

- Soy protein. The protein left over when you have taken the oil out of soy beans.

- Yeast extracts. Made from waste yeast from industrial fermentations, they have everything necessary for microbial growth.

- Peptones, casein hydrolysates. These are partially digested meat or milk proteins respectively. The proteins used are usually waste material from the food industry — nevertheless this can still be an expensive source of nitrogen.

Food processing using enzymes

One of the major uses of enzymes is in the food industry. The food industry is traditionally conservative, preferring to retain existing processes and materials unless new ones provide overwhelming advantages. Nevertheless, biotechnology has provided a range of enzymes which are used in food processing. Among them are proteases, lipases and a range of amylases and glycosidases (see **Glycosidases, Lipases, Proteases**).

In general, enzymes are used to control food texture, flavour, appearance and, to a certain extent, nutritional value. Amylases are used to break down complex polysaccharides, which form viscous solutions or solid gels and do not have much flavour, to simpler sugars which form more fluid solutions and taste sweet. Proteases are used to tenderize meat proteins, especially collagenase, which breaks down collagen, the major protein in connective tissue such as 'gristle' in meat. A widely used protease is rennin, which breaks down milk proteins and so causes them to coagulate, forming the basis of cheese: fungal rennins are now widely used in cheesemaking. Proteases are also used to clarify beers and condition dough for breadmaking.

These enzymes are often added to food during processing, so the amount of enzyme added and the stage of the process at which it acts can be controlled. These are 'exogenous' enzymes. Food also contains endogenous enzymes, enzymes which are naturally present in the food materials. These are also responsible for the changes in texture, flavour, and appearance of the food as it is processed, but are harder to control. Thus allinase helps to develop the characteristic odour of onions, but also can create a bitter flavour in the same food.

Biotechnologists can assist the development of new food enzymes by finding or engineering enyzmes which fit better with the other processes which the food must undergo, like cooking or canning. These improvements could include making them more stable to acid or heat, or making them easier to remove once they have done their job by, for example, immobilizing them on beads or columns so that they can be separated from liquid food or food ingredients easily.

Genetically engineered rennin was the first enzyme produced by recombinant DNA to be approved for food use: it was cloned by Collaborative Research and marketed by Dow Chemicals. As with pharmaceutical products, in the US the FDA provides a rigorous regulatory gateway to using new enzymes in food, especially genetically engineered enzymes, and approval for a food material in the USA is generally taken as being a clear signal to European authorities that the new ingredient is safe. A much wider range of 'novel' food ingredients is approved for use in the Far East, including Japan, than is found in food in the 'West'.

Freeze-drying

This is a common technique, also called lyophilization, for preserving biomolecules and microorganisms. The sample is frozen, often in a solution containing another material such as lactose or trehalose which acts to stabilize it (and is called the excipient). It is then put into a chamber attached to a vacuum pump and, while the sample is still frozen, the chamber is evacuated. The ice sublimes under vacuum (i.e. turns directly into vapour without melting), and the water vapour is removed and trapped in a 'cold trap'. After a while all the water in the sample will have been removed, and what is left is a dry powder or pellet of material.

Commercial freeze-drying apparatus can control the temperature and pressure of the vacuum chamber very accurately, and can heat up the samples to be freeze-dried during later stages to drive off the last remaining water. However simply connecting a jar containing a frozen sample up to a vacuum pump often suffices for research freeze-drying applications.

Freeze-drying is the standard way of preserving microorganisms for long periods of time. It is also a favourite way of formulating biopharmaceuticals, as these protein drugs are often not very stable in watery solution. A good freeze-dried preparation of a protein is a very light fluffy material which, when water or buffer is added, dissolves almost instantly.

Fusion biopharmaceuticals

Several biopharmaceutical proteins have been developed which are fusion proteins — that is, they are the product of two genes which have been fused together so that the proteins that they code for are joined end-to-end. The advantages of such proteins as drugs can be:

- That they have two complementary or synergistic activities in one molecule. Thus, when the molecule binds to a cell, it does two things at once. To get the same effect with two molecules would need much more of both of them, to increase the chance that both would bind at once to one cell.

- The adverse effects or poor stability of one molecule are offset by the properties of the other.

- One molecule acts as a 'targeting' mechanism to bring the other to the site where it is meant to act.

Examples of such fusion peptides are the CD4–IgG combined molecule which Genentech has developed as a potential AIDS treatment, and the Immunex GM–CSF IL-3 fusion. The CD4–IgG blocks binding of the AIDS virus to cells, and is much more stable in the blood than the free CD4 molecule itself. GM–CSF and IL-3 have synergistic effects of stimulating bone marrow to produce white blood cells, so that linking the two together produces a potentially more powerful compound than the two molecules separately. However neither of these have come near to being launched as a drug yet.

See also **Fusion protein, Immunotoxins**.

Fusion protein

A fusion protein is a protein in which part of the chain of amino acids comes from one protein sequence and some from another. 'Biotechnology' is a fusion word, with the 'bio' of 'biology' fused onto 'technology'. Fusion proteins are produced by splicing the gene for one protein next to or into the gene for another: the genetic apparatus reads the gene fusion as a single gene, and so produces a fusion protein.

Fusion proteins are used in a number of biotechnological applications.

- To add an affinity tag to a protein.

- To produce a peptide as part of a larger protein, which is then cut up after it has been made by cloning.

- To produce a protein with the combined characteristics of two natural proteins (for example, a chimeric antibody).

- To produce a protein where two different activities are physically linked (e.g. enzymes for substrate channelling or as a fusion biopharmaceutical).

In practice many proteins are expressed as fusion proteins during research. It is easier to splice the gene for a potentially interesting protein into the middle of another gene than to get it positioned exactly correctly behind a promoter sequence so as to express it as a protein with no additional amino acids.

See also **Affinity tag, Fusion biopharmaceutical.**

Gas transfer

One of the most important characteristics of a fermentation system is the rate at which gas can be transferred from gas phase into solution. Often the rate at which the organisms in the fermenter can metabolize is limited by how fast they can be provided with oxygen or have carbon dioxide, ammonia, or other 'waste' gases removed. Many fermenter design features are aimed at optimizing this transfer rate.

There are several basic methods. Smaller bubbles of gas have a larger surface area per unit volume and greater internal pressure than larger bubbles, so gas diffuses out of them faster. Therefore the smaller the bubbles you can make, the faster oxygen diffusion will occur. The sparger, the pipe system which delivers gas to the base of a tank fermenter, is responsible for breaking up into bubbles the gas flow into the bioreactor and making sure that they are evenly distributed in the reactor volume.

Other methods ensuring good gas transfer all rely on increasing the surface of liquid in contact with the gas. Bubbling the gas through the liquid spreads the gas out — other methods spread the liquid out, for example in a thin sheet (in a pond), or in a thin permeable tube, as in a **hollow fibre** bioreactor.

Gel electrophoresis

Gel electrophoresis is one of the most common analytical methods in biochemistry and molecular biology. Samples are put at one end of a slab of polymer gel (any jelly-like material). An electric field across the gel pulls the molecules through it — smaller molecules can pass through the gel more easily and so move towards the other end faster. Thus molecules are separated mainly according to size.

GEL ELECTROPHORESIS

A large number of materials are used to make the gel, but by far the most common are agarose (for DNA and RNA) and polyacrylamide (for DNA in DNA sequencing and for proteins). Polyacrylamide gels are often called PAGE gels (polyacrylamide gel electrophoresis). Various chemicals can be included in the gel to help the separation, such as the detergent sodium dodecyl sulphate (SDS) in protein gels which unfolds all the proteins, and urea in DNA sequencing gels which does the same to DNA.

A recent variation in DNA gels is pulsed field gel electrophoresis (PFGE) and orthogonal field gel electrophoresis. These also use electric fields to separate molecules, but with several sets of electrodes: the electric field is switched between them, which encourages the DNA to wiggle its way through the gel matrix, now heading one way, now heading another. This helps the separation of very large DNA molecules — up to the size of whole yeast (but not human) chromosomes.

Variants on gel electrophoresis are isoelectric focusing gels, which separate macromolecules on the basis of their isoelectric point (roughly the number of different charged groups they have) rather than on size. O'Farrel gels run an isoelectric focusing gel down one side of a gel slab, and then do a standard PAGE at right angles along the length: this produces a two-dimensional pattern of protein spots which is as characteristic of a mixture of proteins as a fingerprint.

GEL ELECTROPHORESIS

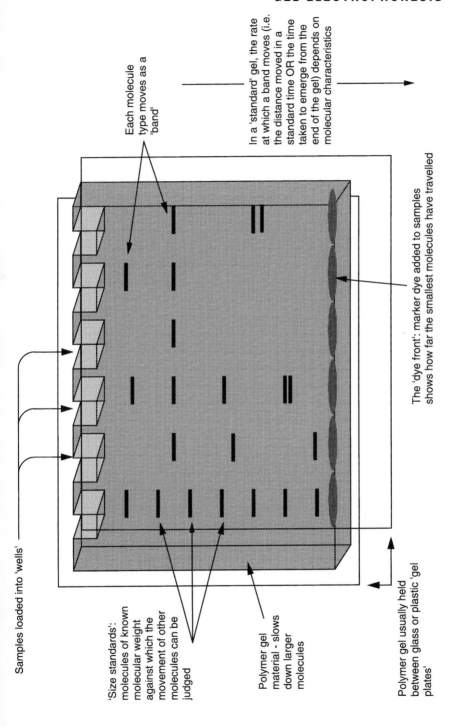

Each molecule type moves as a 'band'

In a 'standard' gel, the rate at which a band moves (i.e. the distance moved in a standard time OR the time taken to emerge from the end of the gel) depends on molecular characteristics

The 'dye front': marker dye added to samples shows how far the smallest molecules have travelled

Samples loaded into 'wells'

'Size standards': molecules of known molecular weight against which the movement of other molecules can be judged

Polymer gel material - slows down larger molecules

Polymer gel usually held between glass or plastic 'gel plates'

Gene

A gene is a section of DNA that codes for a defined biochemical function, usually the production of a protein. DNA (deoxyribonucleic acid) is made up of repeating units which vary in their chemical details (much as a magnetic tape is basically the same all the way along, but varies in the details of its magnetization depending on what has been recorded on it). The parts of the DNA that differ are the bases, so called because they are the chemically basic (alkaline) part of the overall acidic DNA structure. In DNA there are two strands winding round each other in a double helix, so the bases come in pairs. In RNA there is only one strand. Molecular biologists use base and base-pair fairly indiscriminately to mean the length of a piece of DNA or RNA, as RNA is copied off DNA base-for-base in the process of transcription.

Genes are arranged on long DNA molecules called chromosomes, which may contain a few dozen genes in a few tens of kilobases (1 kb = 1000 bases) in the chromosome of a virus, to tens of thousands of genes in hundreds of megabases (1 Mb = 1 000 000 bases) of DNA in the chromosomes of higher plants and animals. All the genes (and hence inevitably all the chromosomes) in an organism make up what is called its genome. The human genome is about 3 billion bases long.

In bacteria, genes that are regulated together (i.e. that are turned on at the same time and by the same stimulus) can be arranged in a tight cluster called an operon. This has one control region at one end, and then a series of coding regions, i.e. regions of DNA that code for single proteins. This whole cluster is transcribed as one RNA, which is then decoded into multiple proteins by the enzymes of the cell. This operon structure is virtually unknown in higher organisms.

So that all genes are not active all the time, genes need control regions attached to them to regulate their activity. In a bacterial operon, these regions are located at one end of the gene. In eukaryotes the control regions (or control elements, as they are usually very short sections of DNA) are mostly at the start of the gene, but can also be spread quite a way away from that start, both within the gene itself and away from it. The key control element, which signals to the RNA polymerase enzyme that there is a gene there at all, is called a promoter — it is essential to have one if a gene is to function. In bacterial systems there may also be an operator, which controls how fast and when the gene is transcribed. In eukaryotic systems there may be an enhancer, or indeed several — these elements boost the transcription of the gene in some circumstances. Both prokaryotic and eukaryotic genes can have a variety of other short sequence 'elements' near their start that allow them to be transcribed or prevent their transcription in the presence of specific materials.

Gene library

A gene library is a collection of gene clones which contains all the DNA which is present in some source, but split up and joined onto suitable vector DNAs. It is also sometimes called a gene bank. If the original source of the DNA was the original DNA from a living organism, then the library seeks to include clones of all that DNA: it is called a genomic gene library, because it contains all the DNA from that organism's genome ('genome' is simply the word for all the genes, or DNA, in an individual). If the DNA is from some other source such as copy-DNA (cDNA) made by enzymatic copying of RNA, then the maker of the library seeks to include representative clones from all that source, and, in this case, would be called a cDNA library.

Gene libraries are not organized like book libraries, and can only claim to be complete because the number of clones in them is sufficiently large for us to be very confident that all the clones we want to be there are there, i.e. that there is a very small chance that anything has been left out. Usually genomic gene libraries are meant to be 95 or 99 per cent complete, so that there is a 95 or 99 per cent chance that the gene you are looking for is in there somewhere.

The number of clones needed to get a complete gene library depends on how big the cloned pieces of DNA are, and how big the genome or mRNA population is. Thus, if you are using a lambda phage vector to make a genomic gene library from human DNA, you need about 500 000 clones. However cosmid cloning vectors can hold substantially more DNA — one would only need 200 000 of these. YAC vectors hold ten times as much DNA again, so one would only need 10 000 of these. This is why people use YAC vectors to make genomic gene libraries, as screening 10 000 clones for the one required is almost invariably easier than screening 500 000

Gene synthesis

This is the complete synthesis of a gene using a DNA synthesizer ('gene machine'), rather than cloning it or assembling it from cloned fragments of DNA. Because most genes are longer than the maximum length of DNA which can be made conveniently on a DNA synthesizer, genes are usually assembled from a number of oligonucleotides. Each section of the gene is hybridized to the next section, and when the whole assembly has been hybridized together, the DNA sections are joined together, enzymatically to make one double helix. This requires the oligonucleotides to be designed carefully, so that they only hybridize to their correct partner and not to other oligonucleotides in the mix.

Other concerns involve making sure that the same sequence is not repeated too often within the gene (as repeated sequences can be targets for rearrangements of the DNA within bacteria), and making sure that the codons used are suitable. Different codons which code for the same amino acid are not used with equal frequency, and generally the more frequently used codons are translated faster than rare codons. However, *which* codon is used most frequently depends on the organism in which the gene is to be expressed.

Other features of the gene, such as the presence or absence of restriction sites, and suitable 'sticky ends' so that the final gene can be cloned into an expression vector easily, are also important.

Gene therapy

Gene therapy is changing the genetic make-up of a human. There are two approaches, germ-line gene therapy and somatic cell gene therapy. The former changes the 'germ cells', the cells which make sperm or ova. This has a permanent effect on all the individuals who are the descendants of whoever had the therapy. Somatic cells are the other cells in the body, i.e. the muscles and bone and nerves etc. Changing these does not affect the germ cells, but does affect the engineered person.

Gene therapy of the germ cells is usually confined to animals, where it is called transgenic technology.

Somatic cell gene therapy can be aimed to correct a genetic or a non-genetic defect. Current therapeutic targets include both categories.

A relatively easy route to somatic gene therapy is bone marrow therapy, as bone marrow is relatively easy to take out and replace, and generates itself inside the body. An engineered stem cell can reproduce itself inside the bone marrow, creating engineered blood cells as it does so. Targets for bone marrow therapy include SCID (severe combined immunodeficiency disease, a very rare genetic disease caused by a deficiency in the enzyme adenosine deaminase, ADA). W. French Anderson and Michael Blaese tried a gene therapy treatment for SCID on a 4-year old girl in late 1991.

Other targets include various types of cancer. Teatments used include the introduction of cells engineered to produce more tumour necrosis factor (TNF) or interleukin (IL-2 or IL-4) into cancer patients, where they are hoped to be able to assist in destroying the cancer.

A version of somatic cell therapy that does not involve genetic engineering at all is autolymphocyte cell therapy (ALT), or autologous gene therapy. This removes a cancer patient's lymphocytes (as opposed to their bone marrow) and uses a combination of cytokine treatments in the laboratory ('*in vitro*') which stimulates them to reject the patient's own cancer.

There have been several proposals for ways of getting DNA into cells while they are still in the patient's body. Routes suggested include:

- Use of retrovirus vectors. Retroviruses efficiently insert their RNA into cells, copy their RNA into DNA, and then insert that DNA into the cell's chromosome. In principle this ability could be harnessed to carry other DNAs into a patient's cells (*see* **Retroviruses**).

- Biolistics. As well as delivering DNA into isolated cells, biolistics can be used to put DNA into cells that are still part of an animal (*see* **Biolistics**).

- Injection. Simply injecting DNA complexed with calcium phosphate into liver or muscle causes some cells to take the DNA up and express the genes in it. This has attracted a lot of attention, because it offers the hope of a gene therapy treatment for muscular dystrophy, one of the most common genetic diseases.

- Using lipsomes. DNA which has been encapsulated into liposomes and injected is taken up by the liver and, to a lesser, extent, by the spleen, and any genes it carries are expressed briefly.

See also **Genoceuticals, Gene therapy — regulation, transfection, transduction, transformation**.

Gene therapy — regulation

Applying gene transfer techniques to humans, usually called gene therapy, has been a major problem for legislators and regulators as well as for scientists. Since the Martin Cline experience in 1980, there has been substantial reluctance to let anyone put genes into anyone else, no matter for what reason. Cline, a UCLA researcher, wanted to put the genes for β globin into patients suffering from thalassemia, a genetic disease caused by defects in the β globin genes. Refused permission to do so in the US, he performed the medical parts of his experiments in Israel and Sardinia (which have much higher incidences of thalassemias). This provoked universal disapproval and a determination to make sure that any future gene therapy experiments were tightly regulated. (The experiment was a complete failure).

Every agency or pressure group with an interest in biomedicine wants to have a say in whether gene therapy can take place. In late 1990 the first gene therapy experiment, giving a SCID patient the gene for ADA (adenosine deaminase), took place. Before it could do so, the experimenters sought approval from the following agencies, any of which could have stopped the experiment.

- National Institute of Health (NIH), Biosafety Committee. Concerned with the technical safety aspects of the experiment.

- National Cancer Institute review board.

- National Heart, Lung and Blood Institute review board. This body and the NCI were funding the experiment.

- NIH's Recombinant DNA Advisory Committee (RAC). This advises on whether any experiment involving recombinant DNA should be allowed. There is an RAC subcommittee on gene therapy, which also had to give its permission.

- The acting director of the NIH.

- Food and Drug Administration (FDA) external advisory committee (as this was an experimental therapeutic procedure).

Although the little girl patient receiving this experimental treatment was released after it was over, making this a formal case of the release of a genetically manipulated organism (GMO) into the environment, the Environmental Protection Agency was not consulted.

Genetic code and protein synthesis

The genetic code is the code used by living cells to turn the information in DNA into the information needed to make protein. How this works is not essential to an understanding of much of biotechnology — the genetic machinery can be treated as a 'black box' for even quite advanced discussions.

The information in DNA is held in the sequence of the four bases of DNA (adenine, guanine, cytosine, and thymidine). This information is transcribed into base sequence in RNA, and then translated into amino acid sequence in protein, the latter occurring on the ribosomes. The RNA is made starting at the 5' end, and is translated starting from that end too:

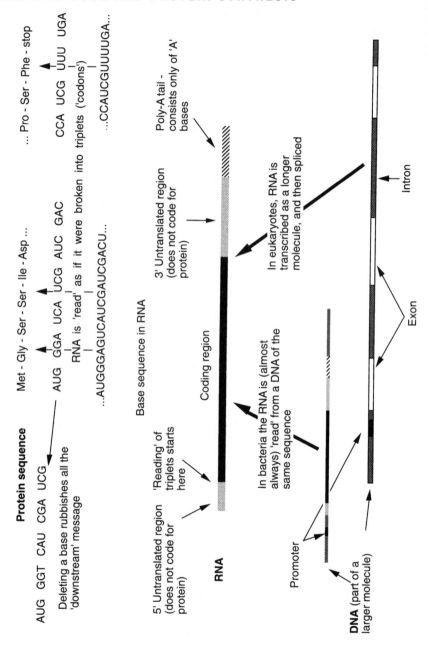

Protein sequence

AUG GGT CAU CGA UCG

... Pro - Ser - Phe - stop

Met - Gly - Ser - Ser - Ile - Asp ...

Deleting a base rubbishes all the 'downstream' message

AUG GGA UCA UCG AUC GAC   CCA UCG UUU UGA

RNA is 'read' as if it were broken into triplets ('codons')

...AUGGGAGUCAUCGAUCGACU...   ...CCAUCGUUUUGA...

Base sequence in RNA

5' Untranslated region (does not code for protein)

'Reading' of triplets starts here

Coding region

3' Untranslated region (does not code for protein)

Poly-A tail - consists only of 'A' bases

In eukaryotes, RNA is transcribed as a longer molecule, and then spliced

RNA

In bacteria the RNA is (almost always) 'read' from a DNA of the same sequence

Promoter

DNA (part of a larger molecule)

Intron

Exon

protein is made starting from the amino end (N-terminus). The sequence which codes for a protein starts with the three-base sequence AUG or (less commonly) GUG, followed by a sequence of bases which is read as triplets, called codons. Of the 64 possible triplets 61 code for a unique amino acid, the remaining 3 being termination codons (i.e. they code for 'stop').

Because there are 20 amino acids and 64 triplets, some amino acids are coded by more than one codon. Once the start code has been found, the cell reads off the other triplets starting from the AUG (or GUG). The way in which the cell reads the message off is called the 'reading frame', as if the cell laid a frame of squares three bases long over the RNA and read what was in each box. Clearly the loss of a single base would then throw out all the cell's reading of the subsequent triplets. Such a mutation is called a nonsense mutation, because it makes a nonsense of the rest of the protein.

Although most of the code is shared between all living things, there are some differences: mitochondria, which have some of their own DNA, do not have quite the same genetic code as the cells in which they reside, for example.

In addition, the DNA sequence (and hence the original RNA sequence) need not be the same as the sequence that is eventually translated. There is a substantial amount of editing of RNA. Segments called introns (which occur in most eukaryotic genes) whose function is unknown are removed, a process called splicing. In some eukaryotes extra uracils are added into specific sites inside the RNA in a process called RNA editing. There are even a couple of known cases of joining bits of different RNA molecules together, known as trans-splicing.

These complications have two implications for biotechnologists. Firstly, it is often not possible to express a eukaryotic gene in a prokaryote. Even if prokaryotic promoter sequences are spliced on, the prokaryote will be incapable of carrying out eukaryotic post-transcriptional modification to the RNA to make it readable. For this reason, many protein expression projects prefer to start with a cDNA clone (a cloned DNA made by enzymatic copying of the final RNA) rather than the original gene. Secondly, although sequencing DNA is easier than sequencing protein, it is not always safe to extrapolate from the DNA sequence to the protein it may code, because of variations in post-transcriptional modification of the RNA and variations in the genetic code.

Genetic disease diagnosis

A genetic disease is one which is caused by a gene, so we 'inherit' the disease from our parents. For a true genetic disease anyone with the correct genotype (collection of genes) will display the phenotype (the physical manifestations of the genes). In practice quite a lot of genetic diseases have incomplete penetrance: that is, the genes do not always cause the effect they are meant to. This makes detecting them rather complex.

Molecular genetics has made huge advances in medical genetics, particularly through making available DNA probes to detect the genes which 'cause' genetic diseases even when they are not causing them — for example when a gene is present in a carrier, or when a dominant allele which causes a disease late in life is present in an infant. These probes have been used both to identify the gene and to diagnose carrier status in people who carry the gene but do not have the disease.

Identifying the gene can be done in two ways. The 'traditional' way is to know how the disease is caused, and hence what protein defect causes it, and so clone the gene from knowledge of the protein. The 'reverse genetics' approach is to use the gene probes to localise the gene whose defective allele causes the disease to a particular chromosome, an approach also called positional gene cloning. This is usually done by linkage analysis. The gene itself may be cloned by one of a variety of methods such as chromosome walking or chromosome jumping. Essentially these are ways of using a piece of DNA which you have cloned to identify bits of DNA from nearby spots in the chromosome.

The genetic diseases for which cloned probes (i.e. probes which identify the gene itself) have been isolated include the haemophilias, thalassemias, sickle cell disease, Duchenne and Becker's muscular dystrophy, retinoblastoma, and cystic fibrosis. A large number of cloned probes which detect loci closely linked to other disease genes, and hence which can be used for medical genetic diagnosis, have also been cloned.

See also **Predisposition analysis, Reverse genetics**.

Genetic engineering

A general term for the directed manipulation of genes, and usually used synonymously with genetic manipulation or genetic modification. A wide range of technologies are involved in this, but most involve the recombinant DNA techniques

Genetic engineering falls into several different categories depending on what is being engineered.

- Bacteria, yeast. This is 'traditional' genetic engineering (i.e. over 10 years old). Using recombinant DNA techniques, genes are put into microorganisms to make them produce something we want, be it insulin or better beer or protein for food.

- Animals. Genetically engineered animals are usually called '**transgenic animals**'. They are produced by a combination of *in vitro* fertilization techniques (IVF) and recombinant DNA technology, and produce animals which pass on their genetic modification to their offspring: they have a germ line modification.

- Plants. Genetically engineered plants are also sometimes called transgenic plants. They are created through the use of plant cloning technologies, which involve growing plants from isolated plant cells.

- Humans. Although the genetic engineering methods applicable to cows or mice are, in theory, applicable to humans, they have not been applied for obvious ethical reasons. Some experiments treating disease have been performed: these do not modify the germ cells, only the 'somatic cells'. This is generally called **gene therapy** or somatic cell therapy, rather than the more spectacular (and potentially publicly alarming) term 'genetic engineering'.

See **Embryo technology, Recombinant DNA**.

Genetic information

Projects such as the human **genome project** and the development of tests for genetic predisposition to disease have led to a lot of talk about how human genetic information could or should be used. This contrasts with genetic information about animals, plants, or microorganisms, which is not thought to have the same ethical status: debates about who 'owns' the human genome focus on high ethical philosophy, those about pig genomes centre around the patent courts.

Several countries have suggested legislation on the use of human genetic information, with DNA-based methods particularly in mind. Denmark resolved to introduce legislation banning the use of genetic information for insurance, pension, and employment purposes in 1991. In the US, California, Texas, and Oregon have introduced similar measures, and New York has a scheme for regulating genetic testing laboratories. The US also has a Genetic Information Act, which prevents the use of genetic information in hiring federal employees.

So far, no one has resolved the problems of copyright and the ownership of proprietary 'DNA probes' in human genetics. Indeed, this is probably one of the most problematic areas of the regulation of applied recombinant DNA work. This is in part because of a confusion with the debate over abortion, in part because of the history of the 'eugenics' movement in Europe (and not only in Germany, by any means, although Germany is particularly sensitive on this issue). Also, as has been the case for almost every biotechnology development since 1970, there is a general belief that 'it will never really happen'. As human genetic testing is already widespread, this is perhaps not very far-sighted.

Genoceuticals

A vogue term for a version of gene therapy where a gene is placed into a cell and there produces a pharmaceutically active protein. So far, several studies have shown that DNA can be placed into the cells of adult mice and rabbits, and that the DNA can work there to produce proteins. This has two potential applications: however both are speculative at the moment, not having been tried properly even in animals.

'Genetic antibiotics' are genes which have some anti bacterial or (more usually) anti viral activity. The genes are placed into the cells which are the potential targets of the parasite. For example, a gene for a toxin could be linked to a controller gene which is activated by a virus: when the virus infects the cell the toxin gene is turned on, toxin is produced and the cell dies.

The other application is to insert genes which will themselves make biopharmaceuticals. For example, calcitonin has been suggested as a treatment for osteoporosis, a bone-wasting disease suffered most frequently by older women. However calcitonin is a protein, and difficult to get into the body: consequently it has to be injected frequently. A genoceutical approach to osteoporosis would be to '**transfect**' the gene for calcitonin into some suitable cells in the individual: these would then produce the hormone steadily for weeks or months.

The reason that this has not been tried so far include technical barriers (it is very hard to get genes into adults reliably and reproducibly), potential problems with side-effects (the genes you put in might disrupt an oncogene, triggering a cancer — and this only need happen in one cell), and the considerable social concern over the use of gene therapy for any application.

Genome project (HUGO)

A genome project (and despite talk about 'the' genome project, there are several competing ones) is a project to determine the exact genetic structure of an organism's genome. This is usually taken to mean the sequence of all of its DNA.

The Human Genome Project is a project to determine 'the' base sequence of all the DNA in humans. It is under the international umbrella of the Human Genome Organisation (HUGO), and is funded primarily by the Department of Energy (DOE) and National Institutes of Health (NIH) in the US and the European Commission (EC) in Europe.

The project started largely because molecular biologists realized that they could sequence the whole human genome, given the money. It is supported strongly by the biotechnology and pharmaceutical industries because it will provide a database of information from which companies can obtain the DNA sequence, and hence protein sequence, of all the proteins in humans, including those which are the potential targets of new drugs, and because it could be of substantial assistance in medical genetics, including the diagnosis of inherited predisposition to disease.

In order to make the sequencing of the three billion bases of DNA in a human genome feasible, the genome projects have set up a number of less ambitious milestones along the way. The first is a complete genetic map of man, defined by **RFLP**s. The second (which looks like the one that will be done first) is a complete sequence of all the cDNA in man. In any case, it is unlikely that the human genome will be sequenced indiscriminately: some bits are much more interesting than others.

As well as the human genome project(s), there are genome projects for pig, the fruit fly *Drosophila*, the weed *Arabidopsis thalliana*, the microscopic worm *Caenorhabdis*, yeast, and *E. coli*. The yeast and *E. coli* projects are likely to be completed in the next decade, as it is believed that nearly all the DNA in these small organisms is essential for their survival, and hence is biologically interesting. By contrast, some scientists believe that over 90 per cent of human DNA is essentially junk.

GLP/GMP

These stand for Good Laboratory Practice and Good Manufacturing Practice. They are a code of practice which is designed to reduce to a minimum the chance of accidents which could affect a research project or a manufactured product.

The GLP and GMP prescriptions are quite voluminous, but boil down to a few key points. The essential point of both GLP and GMP is that everything is recorded, and only established procedures are used by people who have been trained to use them. This may seem obvious, but extends to *everything*: in a true GLP laboratory, for example, only staff who have been trained to use a balance may use it; every weighing has to be checked by another person (who has also been trained for *that specific* balance) who has to sign to say that they have checked that the weight of material is correct; weighing has to be done according to a written standard operating procedure (SOP) for using *that* balance; the protocol used has to be noted in the record of the experiment and so on. All records are kept, and have to be archived on microfiche or magnetic tape. Samples of every batch of material that is used in the experiment or the manufacturing process also have to be archived, so that they can be referred to in future should this become necessary.

Using procedures of this type, it is possible to trace exactly who did what at every stage of an experiment or a manufacturing process. Thus, if there is a problem afterwards, the GLP or GMP user can either point to a specific material or standard operating procedure which caused the problem, or can point to exhaustive documentation which shows that the problem is not their fault. This can be very important in pharmaceutical development and manufacture (GLP was set up after some severe side-effects of a drug were overlooked during preclinical research because the protocols for an experiment were faulty). Many biotechnology companies claim to work to GLP or GMP (depending on whether they are concerned with R&D or manufacturing). In practice, many who claim to do so do not actually work to GLP. It is enormously difficult to do innovative research to GLP, where you have to define a set of standard operating procedures, train the staff formally etc. just to do one experiment which might take half a day. GLP is more relevant to pharmaceutical development (where a large number of very similar experiments are performed). GMP is a standard requirement for pharmaceutical production, and for a number of other industries.

GMP also stands for Good Microbiological Practice, a code of laboratory practice for performing basic microbiology safely. In this sense, GMP is simply a way of minimizing the chance of contamination problems (contamination of the sample or of the laboratory) during a microbiological experiment.

Glucose isomerase and invertase

Glucose isomerase is probably produced in larger amounts for industrial use than any other single enzyme (although by far the largest category of enzymes is the general class of alkaline proteases, used in detergents). It catalyses the interconversion of the two sugars, glucose and fructose. As fructose is chemically more stable than glucose, a mixture of glucose and fructose with the enzyme will end up almost entirely as fructose. This is valuable for the food industry, as fructose is substantially sweeter than glucose, and so you get more sweetness per gram by using fructose.

The usual use for glucose isomerase is to take glucose made by hydrolysis of corn starch and turn it into a mixture of mostly fructose with some glucose. The corn starch is broken down using amylases. The result is called 'high fructose corn syrup (HFCS)'.

Invertase takes sucrose ('sugar') and turns it into glucose and fructose. Thus in conjunction with glucose isomerase it can convert sucrose into HFCS. Invertase can also be used in its own right to convert the easily crystallized sucrose into the less easily crystallized glucose–fructose mixture. 'After Eight Mints', for example, have invertase in their centres — it turns the hard sucrose core (which the chocolate coat was poured onto) into the soft centre which we finally eat.

Glue

Biological glue is one of the many areas where biotechnology and medicine can overlap. Doctors are always interested in new medical techniques for repairing wounds. An obvious way is glue: however the glue must have unusual properties. It must be able to set (cure) in a wet environment, not be broken down by watery liquids, not irritate or poison the body, not induce an immune or allergic response, and the body must be able to break it down after a time if its function is only temporary, like stitches.

The most widely used and discussed glue is the protein fibrin. The body itself produces fibrin, a component of the clotting proteins in blood: however it is not a very strong glue, and, unless derived from human blood (with its concomitant risk of viral contamination), causes a strong immune response. On the other hand, it is a natural human product, and is used in several commercial medical glue applications.

Several marine organisms produce glues which could fit these criteria. Mussels and barnacles produce protein-based glues which could, in principle, be produced by more convenient organisms using biotechnology. Genex has produced a yeast which makes the protein (which has a very unusual amino acid composition, making it difficult for the yeast cell to make it efficiently). The protein also needs extensive and rather specific post-translational modification, which the yeast cannot perform. Thus these proteins are some way from commercialization yet.

Many other organisms make materials which glue them onto things, or things (like eggs or nest material) onto other things. However these have not been investigated nearly well enough to make them attractive for development as medical glues.

Glycation

Glycation is the non-enzymatic reaction of sugars with proteins. Many proteins are glycosylated deliberately by the body, and there are enzyme mechanisms to make this happen. However sugars can also react with the amino groups in proteins in an uncontrolled, chemical reaction. As every part of mammalian bodies have sugar in them, this means that all proteins get glycated after a while.

This is much accelerated by very high sugar levels, or by heating. Hence chemical glycosylation is important to protein processing, and hence flavour formation, in food. Chemical glycation is also very important in the damage done to diabetics when their sugar levels rise above normal, and to all of us as we age. Indeed, one school of thought holds that much of the damage that we know as ageing is due to the effects of glycation. Particularly, glycated proteins can continue to react and form complex, stable crosslinks with sugars and through them with other proteins. These complexes are called advanced glycosylation end-products — AGEs. The body seems unable to remove them specifically, and so they accumulate, crosslinking collagen into a rigid, inflexible net, damaging critical proteins in long-lived nerve cells, maybe even directly mutating DNA.

Glycobiology

Glycobiology is the study of sugars and their role in biology. Usually this is taken to mean the study of complex sugars and what their functions are, and not the metabolism of how sugars are put together and taken apart.

The twin thrusts of glycobiology are the study of glycoproteins, which are proteins with sugar residues attached, and the study of drugs that interact with sugars and affect sugar metabolism, especially the synthesis of those glycoproteins (i.e. glycosylation). Some glycoproteins have as much sugar in them as protein by weight, and the effects of this sugar on the protein can be substantial. Current theory suggests that the sugars on glycoproteins help in protein–protein binding (important for the mechanism by which cells recognize each other and by which viruses bind to and gain entrance to cells).

From this, glycobiology is interested in how the complex sugars on their own interact with glycoproteins, glycolipids (lipids with sugars attached) and each other.

In living systems, sugars, both as simple sugars and as blocks of sugar residues, are joined onto proteins at specific amino acid sites by glycosyltransferase enzymes (a process called glycosylation). Glycolipids can also be joined onto proteins by specific enzymes, a process called glypiation, producing glycolipoproteins. These complex entities are an important part of the surface membrane of cells, and so may be the 'docking molecules' which viruses use to attack cells: consequently biotechnology researchers are interested in them because study of them may lead to better antiviral drugs, and to markers for aberrant cells such as cancer cells.

The application of glycobiology is sometimes called glycotechnology, to distinguish it from biotechnology, a discipline which concentrates more on nucleic acids and proteins. Companies such as Oxford Glycosystems and Glycomed have been set up to exploit the potential of glycobiology. Carbohydrate-based drugs are a popular goal. Thus Oxford Glycosystems are developing a carbohydrate-based anti-AIDS drug (which acts by blocking HIV's mechanism for latching onto cells as it infects them), and Glycomed has a drug aimed at blocking the effect of the glycosylated endothelial-lymphocyte adhesion molecules (ELAMs). Other uses of glycobiological expertise are in manipulating glycosylation in expression systems, and in analysis of carbohydrates and glycoproteins.

See also **ICAM**.

Glycosidases

A group of enzymes which break up complex sugars (such as starch or sucrose) into simple ones (such as glucose or fructose). About 12 000 tonnes of glycosidases are made per annum, almost exclusively for use in the food industry.

The major glycosidase enzymes are amylases (which break down starch) and glucose isomerase (which is used to turn glucose into the sweeter fructose). Amylases break the long chains of starch molecules and similar polymers into shorter segments, and ultimately into glucose. Amylases are commonly extracted from barley, beans, potatoes, and from a variety of fungi.

Other enzymes produced from bacteria and fungi for polysaccharide breakdown are isoamylases and pullulanases. These break off the side branches of starch, and are sometimes called 'debranching enzymes' for this reason. As molecules which are single, unbranched strings of units have quite different rheology from molecules which are branched like a tree, the debranching enzymes can be valuable to the food industry to alter the flow properties or 'mouth feel' of food.

A third group of these enzymes are the cellulases, which break down cellulose. As cellulose is probably the most common biological material in the world, using it as a raw material makes economic sense. However it is very difficult to break down into its glucose monomer units.

Glycosylation (glycoprotein)

Glycosylation is adding sugar molecules onto other things, almost always other molecules and usually proteins: glycosylated proteins are called glycoproteins. Most of the proteins present on the surface of cells, viruses, and in the blood in animals are glycosylated, and so it is considered likely that successful biopharmaceuticals will also have to be glycosylated. Bacteria do not glycosylate their proteins (or rather have quite different types of peptide–sugar linkages from animals), and so genetic engineering techniques have been developed for yeast and eukaryotic cells which *do* glycosylate. Of course, they do not always glycosylate in exactly the same way as human cells do. It is not clear whether many of the peptides produced for biopharmaceuticals actually are more effective or more stable in the body if they are glycosylated.

Sugars can be linked onto the proteins through the amide groups of asparagine in the short peptide sequence Asn–X–Ser/Thr, or, more rarely, through the hydroxyl of serine and threonine. This means that how much a protein may be glycosylated can be predicted to an extent from its amino acid sequence, and hence from the sequence of its gene. Whether this has practical implications, as opposed to being a *post hoc* rationalization of the sugars you find on the real protein, is still debated.

Such glycosylation is a form of post-translational modification, i.e. modification of the protein's chemistry after the protein has been 'translated' from RNA. Other protein glycosylation is chemical, and occurs whenever a protein sits in sugar solutions for a long time. This is also called glycation.

Other molecules can be glycosylated, especially cell surface lipids. The resulting glycolipids are important as tags to allow the body to recognize its cells, especially cells in the blood. Thus they may be important functional components of liposomes, enabling the maker of liposomes to fool the body into thinking that they are cells. Proteins can also have lipids linked on (forming lipoproteins) or even glycolipids. The results cause very different responses from the immune system than the unmodified protein: however making such complex derivatives is much more difficult than making relatively simple glycoproteins.

Although proteins have well defined places where sugars can be coupled onto them, whether sugars are coupled on, and what sugars are coupled, depends on many things. Among them are the cells the proteins are made in, and the metabolic state of the cells. Thus proteins come in variants with different sugars linked onto the same polypeptide chain — these variants are called glycoforms. One cell can also make a mixture of different glycoforms. The different glycoforms have detectably different functional properties in many cases, and are 'seen' as different by the immune system. Viruses in particular come as a population of different glycoforms and not as a single chemical entity: thus HIV (the 'AIDS virus') has different sugar moieties on its surface depending on the cells it is grown in, and exactly what strain of virus is grown in them. These variations certainly bind anti-

HIV antibodies differently, and may affect the immune system of an HIV-positive person differently.

See also: **glycation**.

Gold and uranium extraction

Gold and uranium are mined in commercial quantities using microbial leaching methods. Unlike other metal extractions using bacteria, gold and uranium are extracted using bacteria because of the high added value of the metals and some specific features of the elements.

Gold is usually found as metallic gold mixed with other materials. Crushing the minerals releases the gold metal, which can be separated physically, often by washing. However, substantial sources of gold are ores in which the gold is extremely finely divided and so cannot be released by conventional crushing or milling, called refractory ores. Many such ore types with widely differing chemistry can contain gold, but it is often associated with sulphides, especially pyrites and arsenopyrites, both of which can be oxidized by bacteria. To release the metal, the sulphide must be chemically removed. Bioleaching methods digest the refractory gold ore in a tank fermentation system with a bacterium, usually *Thiobaccilus ferrooxidans*, which oxidizes the sulphide to sulphate. This is usually soluble, so the gold particles are released for mechanical collection. Gold extraction using biological processing is gaining support because the alternatives — oxidation of the sulphur to sulphur dioxide or dissolving the gold out of the mineral using cyanide — are increasingly being considered environmentally unacceptable.

Uranium mining follows more conventional bioleaching lines, with ores that are low in available uranium being incubated with an oxidizing bacterium to release the metal. The tetravalent insoluble uranium is oxidized by ferric ions (generated by the bacteria) or directly by the bacteria themselves to soluble uranium(VI) ions. These can then be recovered from the nutrient mix running off the ore heap.

See also **Leaching**.

GRAS

Stands for 'generally regarded as safe', and is an important category for acceptance of biotechnological products in Western countries, and especially the US.

For microbial or genetically engineered products, regulatory approval for general release is much easier if the product is made in an organism which is GRAS, as the only unknown is then the new product, not the organism as well. For isolated materials, to be accepted as GRAS in one application (e.g. as a foodstuff) can greatly help getting approval for another application (e.g. cosmetics). The exception is usually any pharmaceutical application, where any new product, even if it is believed to be chemically identical to a previous product but made in a new way, must go though the full set of toxicological and clinical trials before it is released.

Growth factors

Growth factors are materials (apparently invariably proteins in mammals) which stimulate growth. They are of great interest as potential drugs (biopharmaceuticals) because they could be used to assist wound healing or even encourage tissue regeneration. The growth factors not only stimulate cell division, but also influence differentiation of the cells and in some cases select which cells in a mixed population divide or differentiate.

The ones most often discussed are:

- Epidermal growth factor (EGF). This stimulates a variety of cells in the upper skin to divide and differentiate. Could have a use in helping wounds to heal.

- Erythropoietin (EPO). A factor which stimulates the cells which give rise to red blood cells. Thus it is used to boost the number of red blood cells in the blood, which is useful for leukaemia or kidney dialysis patients. It is also rumoured to be used by marathon runners to increase their blood's capacity to carry oxygen, a use which is almost certainly dangerous. Erythropoietin is being made for clinical trials by Amgen and Genetics Institute, who are involved in a patent dispute about the protein.

- Fibroblast growth factor (FGF). Stimulates growth of the cells common in connective tissue and the 'basement membrane' which many cells are attached to. It has been suggested as a stimulant to the healing of burns, ulcers, and bones.

- Haemopoietic cell growth factor (HCGF). Stimulates the production of many of the haemopoietic cells, i.e. cells made in the bone marrow and ending up in the blood.

- Neurotropins (*see* **Neurotrophic factor**)

- Platelet-derived growth factor (PDGF). This stimulates connective tissue to grow, and is associated with wound healing.

- Stem cell factor. A protein which stimulates the 'stem cells' from which all the cells in the blood are made. The stem cells reside in the bone marrow. (Actually, many tissues have their own 'stem cells': these are specific for the blood — they are haemopoietic stem cells.)

Hairy root culture

This is a fairly new type of plant culture, which consists of highly branched roots of a plant. A piece of plant tissue (an explant, usually a leaf or leaf section) is sterilized to remove the bacteria on the surface, and then treated with a culture of the bacteria *Agrobacterium rhizogenes*. Like its cousin *Agrobacterium tumefaciens*, *A. rhizogenes* transfers part of its own plasmid DNA to the cells of an infected plant. This causes alterations in the plant metabolism, including alterations in hormone levels. This in turn causes the explant to grow highly branched roots from the sites of infection. The roots branch much more frequently than the usual root system of that plant, and are also covered with a mass of tiny root hairs, hence the name of the culture system.

The hairy root cultures do not require hormones or vitamins to grow, unlike explant cultures or cell cultures of plant cells, so they can grow on simple media of salts and sugars. They are also genetically stable, again unlike explant or cell cultures, so they can be cultured in bulk without the culture changing. Their most significant feature, however, is that they produce secondary metabolites at levels similar to those made in the original plant. Thus they can be used as replacement plants for making such compounds as food flavours or fragrances. As such they are the target of considerable interest and research, although no products have yet been produced

Hairy root cultures have been grown in several large laboratory fermenter systems as well as small pilot plants. They look like a mass of fibres when growing as an unstirred mass: they can be grown in a stirred tank reactor, but they are rather sensitive to being broken up by the stirring machinery. However, because they grow and metabolize more slowly than bacteria and do not have nearly such high oxygen demands, stirring is not usually necessary to obtain a successful culture.

Harvesting

Harvesting in biotechnologcial terms usually means collecting cells or organisms from a growth system. If the cells or organisms are very large (like, say, trout) this is not difficult. However most biotechnology uses single-celled organisms like bacteria or yeast, which have to be actively collected. Among the ways of doing this are:

- Centrifugation. Expensive, but guaranteed to collect any particle. It can be used in small volumes to purify viruses, and anything as large as a bacterium can be processed quite easily.

- Filtration. There are a range of filtration systems. This is cheap and again effective, but usually has limited capacity. The reason for this is that the filter is, of necessity, full of holes that are just a bit smaller than the cells you want to collect. So after a while the cells fill up all the holes, the filter is fouled, and filtration stops. Cross-flow filtration can help to solve this.

- Flocculation. This is popular. By adding an additional reagent to the reaction mix, or by altering conditions, you get the cells to stick together and settle out like snow. This is often the only practical way of removing cells from really big fermentations, and is essentially how yeast is removed from brewing vats after fermentation is complete.

See also **Cross-flow filtration**.

Herbicides and resistance

One of the early targets of genetic engineering used on plants was to make them resistant to common herbicides. If a broad spectrum herbicide was sprayed onto a field planted with such resistant crops, then all the plants except the crop would be killed, thus providing an effective method of weed control without having to develop herbicides specific to each weed type.

The tolerance mechanism has to be designed to fit the herbicide — consequently, different companies have been working on engineering resistance to their particular herbicide. There are two general approaches: alter the enzyme which the herbicide normally attacks so that it is no longer a target for that chemical, or add a system for detoxifying the herbicide in the plant.

There is substantial concern in some quarters about the widespread use of this technology, which is essentially giving the plant kingdom the ability to evade man's most effective herbicides. The concern is that such engineering will lead to increased use of the herbicides, at a time when it is generally accepted that the use of chemicals should be kept as low as possible, and that there is the possibility that resistant crop plants will ' escape' to become weeds, or will even transfer their resistant genes to other weed species.

The groups of pesticides which biotechnologists have examined so far are:

- Glyphosate. Marketed by Monsanto as Roundup, this is a very commonly used herbicide which stops amino acid synthesis. Plants resistant to glyphosate have been created by giving them new, resistant enzymes and by selecting resistant cells and cloning them into whole plants. Monsanto is developing glyphosate-resistant cotton plants, expecting them to be ready for general farm use in the mid-1990s.

- Phosphinothricin (PPT). Produced by Hoechst, this herbicide blocks amino acid synthesis. Resistant alfalfa has been created by isolating alfalfa cells resistant to the herbicide and cloning whole plants from them. Plant Genetic Systems have also engineered tobacco and potato to resist Phosphinothricin.

- Sulphonyl ureas. These block amino acid synthesis. Mutant genes from *E. coli* have been put into plants to confer resistance.

- 2,4-Dichlorophenoxyacetic acid. A compound which mimics plant hormones, and so disrupts their growth. Bacterial genes which break it down have been put into plant cells.

- Triazines (Atrazine, Bromoxynil). These disrupt photosynthesis by binding to a protein (the Qb protein) in the chloroplast. Natural mutants which are resistant to triazines have an altered Qb: a resistant crop plant could therefore be made by putting that altered Qb into the crop plant. Getting this altered gene product into the chloroplast is, however, a major problem. Ciba Geigy is working on an alternative route, putting enzymes which detoxify atrazine into several crop plants: because the detoxification enzymes work in the cytoplasm, this may be a simpler route for the genetic engineer.

Hollow fibre

Hollow fibres are tubes of a material that is porous. The tubes are very small, typically having an internal diameter of a fraction of a millimetre, and so their ratio of surface area to internal volume is very large. This has had two types of application.

Firstly, hollow fibres can be used as filters. Because they have a huge surface area, they take much longer to clog up than normal filters. The filters used in artificial kidney machines are often hollow fibre bundles.

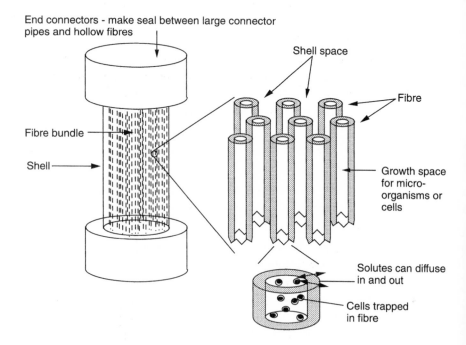

End connectors - make seal between large connector pipes and hollow fibres

Shell space

Fibre

Fibre bundle

Shell

Growth space for micro-organisms or cells

Solutes can diffuse in and out

Cells trapped in fibre

The second use is in the hollow fibre bioreactor. This is a widely used type of bioreactor in which cells are kept inside the hollow, porous fibres, and the culture medium is circulated outside the reactor. The fibres have pores large enough to let the nutrients in and the product out, but not to let the cells out. The fibres are kept inside a shell: the space outside the fibres but inside the shell is called the shell-space.

Hollow fibre bioreactors enjoy very general use in some applications. They are very effective for maintaining mammalian cells in culture because they have a very large surface area for the cells to grow on without needing a large reactor to hold them, and because the nutrient reaching the cells can be kept fresh: mammalian cells can be very susceptible to changes in the medium in which they grow. The reactor also provides an easy way of removing the product that the cells are making: this has meant that hollow fibre reactors have been particularly useful for making large amounts of monoclonal antibodies.

Hollow fibre reactors are less use when the cells themselves have to grow, because it is hard to get at the inside of the fibre to remove surplus cells, and hard to monitor exactly how many cells you have inside the fibres. This has meant that hollow fibre reactors have limited useful for bacterial cultures.

Homologous recombination

Homologous recombination is a biological process by which a living cell joins two pieces of similar DNA together. It is a version of the general genetic process of recombination, by which any two pieces of DNA are spliced together in a living cell. Recombination occurs in all living things: recombinant DNA technology is so called because of the similarity of its gene-splicing technology to natural recombination processes.

Homologous recombination is recombination between two pieces of DNA which are almost exactly the same — i.e. they are 'homologous'. This occurs much more readily than recombination between DNA which is completely different. This is particularly true of bacteria and yeasts.

HOMOLOGOUS RECOMBINATION

Homologous recombination is notoriously much more difficult to achieve in higher organisms such as plants or animals. It is used as a mechanism for ensuring that a cloned gene which the experimenter wishes to put into the chromosomes of a cell is inserted into those chromosomes at a specific point (that is, at the point where the cell's DNA is the same as the cloned DNA). Because of this use, homologous recombination is sometimes called gene targeting. Homologous recombination is used in biotechnology in three areas.

In generating new mutants of many organisms, but particularly yeast, homologous recombination is a method for targeting a specific bit of DNA. A piece of yeast DNA is joined into your **plasmid**: when the whole is introduced into a yeast cell, the piece of yeast DNA can undergo homologous recombination with the corresponding bit of DNA in the original yeast's DNA. The two are spliced together and, because the plasmid is all one piece, this means that all the other bits of DNA in the plasmid are also joined into the yeast DNA. This can be used to join a plasmid into the yeast chromosomes, or, if the yeast DNA is from a known gene, to disrupt that gene by putting a large piece of plasmid in the middle of it.

The second role is in manipulating large plasmids such as the Ti

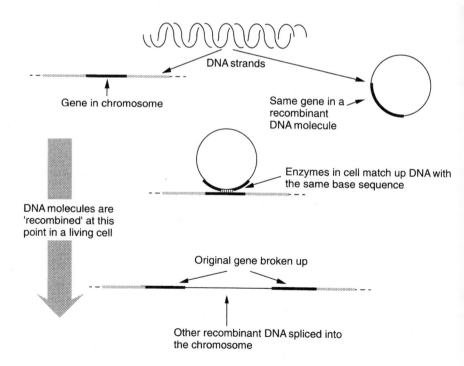

DNA strands

Gene in chromosome

Same gene in a recombinant DNA molecule

Enzymes in cell match up DNA with the same base sequence

DNA molecules are 'recombined' at this point in a living cell

Original gene broken up

Other recombinant DNA spliced into the chromosome

plasmid of *Agrobacterium tumefaciens*, which are too big to alter using recombinant DNA techniques. Genes can be spliced into them in exactly the same way as they are spliced into the yeast chromosome.

The third application is in making transgenic animals (and potentially in gene therapy). Here, again, homologous recombination is used to carry a foreign gene into a cell's chromosome. However the reason for doing it in this case is to avoid disrupting any genes in the target cell, and to make sure that the foreign gene arrives in a suitable chromosomal environment. The DNA which surrounds genes in mammalian cells (and many other types of cells) affects how that gene will be expressed. Thus it is important to target a foreign gene to a suitable place in the host cell's chromosomes, so that the gene functions correctly, and essential that the gene is not targeted to a site where it will damage the functioning of other genes. Homologous recombination offers a route to doing this, and hence making the production of transgenic animals more reliable. It also offers the possibility of useful gene therapy in man, as one of the main problems with gene therapy ideas at the moment is the threat that the 'therapeutic' gene introduced into a patient's cells will do as much damage as the original disease.

Human growth hormone

Human growth hormone (hGH) was one of the earliest proteins to be made by genetic engineering to gain approval for use as a drug: Genentech sold US$150 million worth in 1990. Mammalian growth hormones are produced naturally by the pituitary gland in young animals before and during adolescence. They increase the rate of growth and stimulate the body to put on muscle mass. After the age of 30 production of growth hormone falls: injections after this age cause muscle to build up and fat to decrease.

Human growth hormone is used medically for rare children's diseases where the body does not produce its own growth hormone. It can also be used to treat several diseases where extremely short stature is part of the disease, although not because of a shortage of the growth hormone, such as the chromosomal abnormality Turner's syndrome.

HUMAN GROWTH HORMONE

Recent work suggests that hGH reduces, maybe even reverses, the reduction in muscle mass which occurs with ageing, and also improves skin elasticity and muscle tone. It could therefore be an anti-ageing drug. This has given rise to substantial interest, especially among older bankers, but will be very difficult to prove. Even if it only reduces the effects of ageing, however, without lengthening lifespan, it could still be very attractive. Against this must be set the near certainty that the drug will have some side-effects: whether they will be trivial or life-threatening remains to be seen. There is considerable debate over how trials to test the potential anti-ageing effects of HGH could be carried out: unless ageing is defined as a disease, there is no route for a powerful drug to be tested to treat it. If it is a disease, then the drug must be proven to have some significant effect on it, which could take decades.

A related area of use could be as an anti-wasting agent in a disease such as AIDS.

A third area of use for hGH is simply illegal, but may go on anyway. This is in drugs abuse in sport.

See also **Sports and biotechnology**.

Hybridization

Hybridization has several meanings in biotechnology and molecular biology.

DNA hybridization. This is the formation of a double helix of DNA from two DNA strands. The two separate strands of DNA will come together to make a double helix if their bases are complementary, so that wherever there is an A (adenine) in one strand there is a T (thymidine) in the other, and wherever there is a G (guanine) in one strand there is a C (cytosine) in the other. (In fact there is a slight degree of laxity about this, and, depending on how long the DNA strands are, up to 10 per cent wrong or 'mismatched' bases can be tolerated.) DNA hybridization is used as a method for using one bit of DNA (the probe) to find out

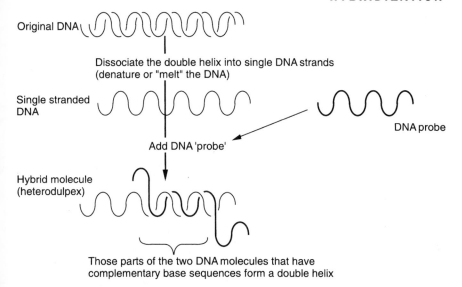

Original DNA

Dissociate the double helix into single DNA strands
(denature or "melt" the DNA)

Single stranded
DNA

DNA probe

Add DNA 'probe'

Hybrid molecule
(heterodulpex)

Those parts of the two DNA molecules that have
complementary base sequences form a double helix

whether a complementary bit of DNA is present in some mixture of DNA species. It is used in 'blot' techniques, **PCR**, **gene library** screening, **DNA fingerprinting** and a range of other techniques.

Molecular hybridization. This is a method for forming a new molecule that has functional parts from two different molecules. It could therefore have a combination of properties from its two 'parents'. Examples of this approach are the new antibiotics that can be made by combining the enzymes that make two old antibiotics in one cell, and making fusion proteins by joining two functional domains of other proteins together.

Cellular hybridization. This is essentially another term for cell fusion.

Species hybridization. This is the formation of a hybrid between two species. Hybrids between related species ('interspecific hybrids') occur fairly commonly in nature. They can be formed between closely related species by suitable breeding programmes: however many species do not hybridize readily, and, apart from a few closely related species such as donkey and horse, animals rarely form hybrids in this way. Alternative methods include making chimeras, cell fusion (for plants — this rarely works for animals) to produce a new species with all the genes of both the original species, or using bacterial plasmids to shuttle genes between bacterial species.

See also **Cell fusion, Chimera, Fusion protein**.

Hydrophobicity

A hydrophiobic molecule is a molecule that dissolves very poorly if at all in water, but which dissolves quite well in a solvent such as butanol or toluene. They are non-polar molecules, i.e. they are essentially electrically neutral all over. The opposite is the hydrophilic molecule, which dissolves well in water or DMSO (dimethyl sulphoxide) but poorly, if at all, in toluene or long-chain alchohols. These molecules usually have partly charged groups on their surface, and often form ions when dissolved in water. Most biological molecules are to some degree hydrophilic, a major exception being the triglycerides ('fats'), which are hydrophobic. Hydrophobic molecules are therefore sometimes called 'lipophilic'.

When given a choice of environments — for example, a mixture of water and oil to dissolve in, hydrophobic molecules will chose a hydrophobic environment (in this case oil), hydrophilic molecules a hydrophilic environment (water).

However there are degrees of hydrophiobicity and hydrophilicity. Thus, among the amino acids, glutamic acid and lysine are highly hydrophilic, because they form ions easily and dissolve readily in water, while tryptophan has a large uncharged side-chain and is more hydrophobic in character. These differences in hydrophobicity can be used to separate the molecules. Hydrophobic chromatography uses this phenomenon: a mix of molecules is passed over a solid material that is mostly hydrophobic in character. The molecules that are more hydrophobic will stick to the material more strongly, and so will not be washed across the solid support as fast as the hydrophilic molecules.

Many biological molecules are sufficiently large to have distinct hydrophobic and hydrophilic bits. These molecules are called amphipathic. If the two regions of the molecule are at opposite ends, then the result is a surface active material: it will tend to congregate at the junction between a hydrophobic and a hydrophilic solvent. Phospholipids are of this type. Phospholipid membranes are arranged so that the 'tails' of the phospholipids form a layer of hydrophobic 'liquid' that dissolves quite different chemicals to the watery phase around it. Proteins also almost invariably have a mixture of 'hydrophobic' and 'hydrophilic' amino acids. The protein folds so that most of the hydrophilic amino acids are exposed to the waterry solution in which it is dissolved, and most of the hydrophobic

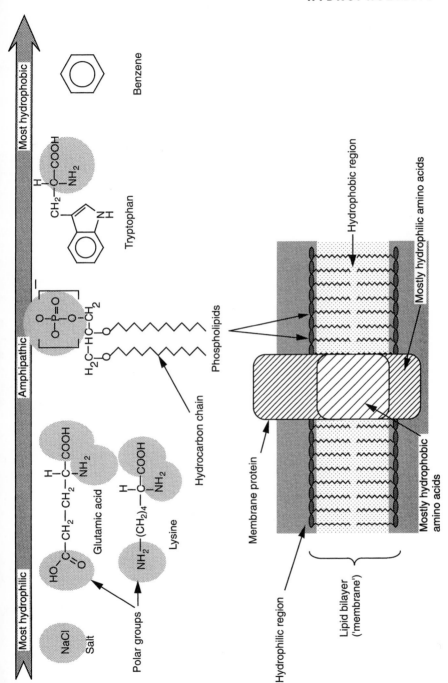

amino acids are tucked away inside the protein. Thus the distribution of hydrophobic and hydrophilic molecules along a protein (sometimes called a 'hydrophobicity plot') can be a clue as to how the protein will fold up. In particular, proteins with a large region of hydrophobic amino acids in the middle of their sequence are often associated with membranes, with the hydrophobic amino acids embedded in the hydrophobic layer in the middle of the lipid layer.

ICAM

Intracellular adhesion molecules, also called CAMs (cellular adhesion molecules). These molecules are present on a wide range of human cells, and are part of the mechanism used by cells to recognize each other. They are glycoproteins, and the sugar residues can be crucial to their function: the differences between some blood groups, for example, are the result of variation in the sugar residues on some ICAM molecules.

ICAMs are important to drug companies, and hence to biotechnology companies, because they are the molecules through which inflammatory responses happen. Thus when your finger swells up after a bee sting, it is because of infiltration of the tissues in your finger by white cells which interact with the cells around them through the ICAM signalling system. Thus there is substantial work in cloning the proteins and using them as targets for, or as the basis of, drugs to moderate the inflammatory response.

Related molecules are ELAMs (endothelial-lymphocyte adhesion molecules). These are proteins on the surfaces of lymphocytes and the endothelial cells (flat cells) which line the surfaces of blood vessels. During inflammation, white blood cells leave the blood and invade damaged tissue to engulf any invading organisms. They also release a range of chemicals which cause inflammation of the tissue. The invasion is partly controlled by the ELAMs, which allow lymphocytes to stick to and recognize the endothelial cells. Modulating this interaction, then, is a potential route to controlling inflammatory diseases.

Imaging agents

A range of proteins are being developed as imaging agents or contrast agents. This means that they are for use with the various types of body-scanner. The proteins (usually antibodies) are linked to a chemical group which allows the scanner to see them very easily. The proteins bind to specific tissue types, usually tumours, and so allow the scanner to distinguish those tissues from the surrounding tissue very easily: in the absence of a contrast agent the target cells could look exactly like the surrounding tissue.

Imaging agents can be made for any of the main imaging systems:

- CT scanning (computerized tomography). This technique uses X-rays, and consequently the label tagged onto the antibody is an X-ray opaque substance, usually something made form a 'heavy' metal like gold.

- PET scanning (positron emission tomography). This technique injects a very small amount of a radioisotope into the body, and then tracks where it goes by following the path of the radioactive particles. The favoured isotope to tag onto the antibody for this is technetium, possibly solely because it sounds technical.

- NMR (nuclear magnetic resonance). This exploits the way that the body absorbs microwaves when it is in a strong magnetic field. Chemical groups absorb microwaves differenly depending on what sort of field they are in and what the group is. A wide range of materials can be used as 'contrast agents' for NMR scanning.

- ESR (electron spin resonance). Little used, but of substantial potential, ESR detects 'unpaired' electrons, electrons which turn up in some types of compunds, usually those involved in energy metabolism. This is different from NMR which usually detects water. NMR and ESR use no radiation, and so are gaining favour as diagnostic systems because of the widespread nucleophobia that is especially prominent in the US.

Immobilized cell bioreactors

Many of the plant and animal cells grown by biotechnologists are most effectively handled not as isolated cells but as cells immobilized onto some support material. This helps to shield them against the stirring forces necessary to mix a bioreactor's contents, and makes them easier to move around and separate from the substrate.

There are a wide range of immobilized bioreactors. They fall into two classes.

Membrane bioreactors. These grow the cells on or behind a permeable membrane, which lets the nutrients for the cell through but does not let the cells themselves out. Variations on this theme are the **hollow fibre** reactor, a common way of growing hybridoma cells for making monoclonal antibodies.

Filter or mesh bioreactors. Here the cells are grown on a open mesh of an inert material, which allows the culture medium to flow past it but retains the cells. This is similar in idea to membrane and hollow fibre reactors, but can be much easier to set up, being similar to conventional tower bioreactors with the meshwork replacing the central reactor space.

Others. In other applications, 'immobilized cells' is often taken to mean that the cells are immobilized onto something not all that much larger than they are, such as small nylon or gelatin beads. The reactor can the handle the beads in the same way as granular catalysts are handled in chemical reactions. There are several ways of doing this. Normal bioreactors of all sorts can be adapted to handle the larger particles. This works best if the particles are of neutral density (as are particles made from most biopolymers). Alternatively, if the particles settle out fast, the bioreactor can be a fluidized bed or a solid bed bioreactor. In the former the particles are kept suspended in a dense, fluid mass by liquid pumped through them from the base. The mass behaves like a liquid, even though it is made of solid particles. In the latter the fluid flow is not fast enough to push the particles apart, and so they sit in a bed at the base of the bioreactor with the fluid flowing past them. Packed bed reactors come in various shapes (conical — 'tapered bed reactor', disc-shaped — 'radial flow packed bed') to help fluid flow through them evenly.

Immobilized cell biosensor

These are biosensors (i.e. detector devices which use a biological bit to allow them to detect one thing at a time) which use living cells as their detector system. They are often called microbial biosensors, as usually it is bacterial cells that are exploited.

As with any biosensor, there are two parts to immobilized cell sensors: the immobilized cell (which does the sensing and produces a very weak signal of some sort) and a system that detects and amplifies that weak signal into a signal that the user can 'read'.

The cell used depends on what you want to detect. Some typical examples of analytes ('things to be analysed') are:

- Amino acids (using bacteria which metabolize them).

- Glucose (using almost any cell).

- Toxic chemicals (using any bacterium which is sensitive to the chemicals to be detected).

- Carcinogens (using bacteria which are defective in DNA repair genes).

- BOD (Biological oxygen demand, i.e. how much organic matter there is in waste waters).

- Heavy metals (using metal-resistant bacteria).

- Herbicides (using plant cells or blue-green algae).

- Toxicity (using cultured animal cells).

Only a few of these have been transformed into realistic sensor systems.

The read-out methods can be equally diverse:

- Gas generation/depletion. A favourite one, measuring the amount of oxygen burnt up or carbon dioxide produced by the bacteria. Rather non-specific, as almost anything the bacterium does burns up oxygen and generates CO_2.

- Light production. This uses luminescent bacteria, either ones which are naturally luminescent or ones which have the relevant genes (for luciferase, the light-generating enzyme) genetically engineered into them. Light production is either a measure of general bacterial well-being (for toxicity sensors) or is coupled to the presence of a specific chemical.

- Direct electrochemical coupling. Some groups are working on hooking the electrode directly into the bacterium's own electron transport system. This is a more sophisticated version of measuring oxygen uptake.

Bacterial biosensors are usually much less specific than other biosensors, as bacteria are very diverse and complex things. However they have the substantial advantage that they are very active, and so make a 'signal' which is easier to detect than that produced by antibodies or DNA probes.

Of the few commercial 'biosensor' systems, several are bacterial; bio-sensors: two luminescence-based bacterial biosensors (for toxicity and BOD measurement) are in use in the water industry, for example.

Immortalization

Immortalization of a cell type is its genetic transformation into a cell line which can proliferate indefinitely. Cells taken from a mammal, called primary cells, will divide in culture for 20–60 divisions, but the stop dividing. This is not because they have run out of room or nutrients to grow, but rather because they have become incapable of dividing any more. They usually show characteristic changes in their structure, and reduce the amount of biotechnologically useful product they produce, be it metabolite or protein. This is called cell senescence, and severely limits the manipulations that primary cells can be put through.

To get around this, cells are immortalized — put through a treatment which allows them to overcome senescence and divide indefinitely, keeping whatever differentiated features they had to start with. This is one in a number of ways. Several **oncogenes**, when transfected into a cell, will immortalize the cell. Some genes from oncogene (tumour-causing) viruses can also immortalize cells, notably the T-antigen gene from SV40 virus. A third route is to look for a spontaneous mutant in the cells you wish to immortalize. This is done by growing a large number of the primary cells in culture and simply looking for any that keep on growing when the others have become senescent. The rate at which this happens varies between different organisms — mice seem to generate immortal cell lines much more readily than humans, for example.

The last common route is via cell fusion. If a mortal, primary cell is fused with an immortalized cell line, the result is usually an immortalized cell. This is how the technology of making monoclonal antibodies immortalizes the one lymphocyte which is to make the antibody characterisic of a hybridoma. All the lymphocytes in a sample are fused with a suitable immortalized cell, so they are all then immortalized: the experimenter can then grow them indefinitely in the search for the hybridoma that is producing the antibody that she wants.

See also **Cell fusion, Cell growth, Cell line**.

Immunization

Immunization is the process by which an animal is made to produce an antibody against something. The animal could be a human or a farm animal, in which case the purpose of immunization is to provide that animal with the ability to make that antibody so that they are protected against a disease. Or the animal could be immunized so that we could collect its blood and extract the antibody from it, thus providing us with a source of that antibody. There are a number of steps involved.

- The animal is injected with the antigen, i.e. the material to which we want the antibody to react. If that is a very small molecule (such as a steroid hormone or a short peptide) then it is usually linked to a much larger molecule, such as a protein. Favourite proteins are bovine serum albumin (BSA) and keyhole limpit haemocyanin (KLH).

- If the object is to obtain an antibody (as opposed to protecting the animal) then the antigen can be injected together with an adjuvant, a material which increases the immune response. Common materials are mineral oils and similar complex mixtures that cause inflammation. A common brew is Freunds complete adjuvant.

- Boosters. The first injection will give rise to a primary immune response, the production of a relatively small amount of antibody. The antibody will be mostly IgM (*see* **Antibody structure**) and its K_a will be low. If the same antigen is then injected again, a secondary response will occur, producing much more antibody, this time mostly IgG with a higher affinity. This subsequent injection is called boosting, and is usually done several times.

● Titres. To test how the immunization is going, a small sample of blood is removed and the ability of the antibodies in it to bind to the antigen is tested. The blood is diluted until the antibodies in it are so dilute that they can no longer bind to the antigen to any detectable degree. This dilution is called the antibody's 'titre'. When measuring the strength of an antibody preparation, people often cite a dilution figure — 1 in 100 000 being pretty good, 1 in 1000 being fairly hopeless. This is the dilution they are referring to. As immunization proceeds with additional boosters, the titre of antibody should go up as the amount of antibody and its affinity goes up.

See also **Binding**.

Immunoconjugate

A compound which is a combination of an antibody molecule (or part of one) and another molecule. There are several types.

Immunotoxins (*see* **Imunotoxins**).

Antibody contrast agents and tracers. These are used in conjunction with 'scanners' — CT (computerized tomography, an X-ray technique), PET (positron emission tomography, a radioactive tracer system) or NMR (nuclear magnetic resonance) diagnostic devices. They all produce images of the insides of a patient, but these images can be much improved (in the case of CT and NMR) or are only possible (in the case of PET) if some chemical is injected into the patient which the scanner can detect. If the chemical is linked to an antibody, then the scanner becomes a very sensitive way of tracking where the antibody ends up. Contrast agents are chemicals which increase the 'darkness' of the scanner image, and apply to NMR and CT scanners (and to conventional X-rays, too). Tracers are materials which do something unique, so they 'light up' in the scan: some NMR reagents and PET scanner chemicals fall into this category.

Antibody–enzyme conjugates. These are complexes where the antibody has been chemically linked to an enzyme. They are widely used in **immunoassays**, where the enzyme acts as a 'flag' to mark the presence of the antibody. A few nanograms of antibody can easily be detected if a suitable enzyme is attached. Common ones are horse-radish peroxidase (HRP) and alkaline phosphatase(AP).

See also **Imaging agents**.

Immunodiagnostics immunoassays

One of the success stories of biotechnology, these are medical diagnostic methods which use antibodies. The antibody is used to detect the presence of something in a sample. The antibody latches onto its target very specifically, so it is a very precise reagent. It can also latch on to the antigen at very low concentrations, and so can be a very sensitive test. This combination has meant that, in the ten years since they have been commonly available, monoclonal antibodies have come to be used in about 20 per cent of all medical diagnostic procedures. Exactly the same test technology can be used in other, non-medical applications, which are called immunoassays.

The problem with immunodiagnostics is that the antibody does not do anything obvious when it latches onto its target, so we have to arrange the assay so that some other process detects that binding has occurred. There are various facets to this.

The Label. Antibodies can be labelled in various ways. As well as labels used for imaging agents (*see* **Imaging agents**), immunodiagnostics can use a variety of labels in '*in-vitro*' assays. These usually have different names:

- ELISA — enzyme-linked immunosorbant assay. Uses an enzyme label on the antibody.

	If antigen is there	If there is no antigen
Sandwich assay	Antigen links the labelled antibody onto a solid material	If no antigen present, the label is not bound to the solid support
Competitive assay	Antigen binds to the antibody in solution	If no antigen is present, then the antibody is free to bind onto the solid support

Type of assay	When antigen is present	When antigen is absent
Latex agglutination assay	Microspheres held together by antigen — Latex forms clumps	Microspheres not held together — Latex forms uniform suspension

- RIA — Radio-immunoassay. Uses a radioactive label on the antibody or antigen.

- FIA — Fluorescent immunoassay Uses a fluorescent tag on the antibody or antigen.

The second facet is the chemical format of the assay — which reagent is attached to which object. Common aspects of assay formats are:

- Sandwich assay. Here two antibodies are used which bind to different parts of the antigen. One is trapped ona solid surface (e.g. the bottom of the wells on a 96-well plate, *see* **Standard laboratory equipment**). The other has a label attached to it. If the antigen is present it links the two, and so the label stays in the plate.

- Competitive assay (competition assay). This is like a sandwich assay, but the analyte is a small molecule which competes with the binding of an enzyme, chemically linked to the antigen (producing an antigen–enzyme conjugate). This is virtually the only way of making an immunoassay which can detect a small molecule.

- Latex. 'Latex' particles are very small particles of plastic, usually coated with the antibody: typically they are polystyrene spheres 100 nm–1 μm across. In the presence of antigen the particles stick together into larger lumps, held together by the antibodies which coat them, hence the name latex agglutination assay.

The third facet is the physical format of the assay. Assays can be:

- Homogeneous, i.e. simply giving a result when the sample is added (together with suitable reagents), as a colour pH indicator would.

- Microtitre plate format, i.e. done on microtitre plates (which have to undergo a series of washes between each reaction). Doing it on other surfaces — glass plates, silicon chips, etc. can be essentially the same.

- Microparticle-based, i.e. the antibody is bound to tiny beads and these are moved between solutions by centrifugatin, filtration, or other methods (This is different from a latex agglutination assay, where the particles are the read-out system as well).

There are a range of semi-official trade names for more complex immunoassays (the competition for a good acronym for ones immuno-assay is intense). Among the more common are:

- ARIS. This uses a complex reaction in which the binding of an antibody to a synthetic target prevents glucose oxidase from working. This type of assay is now almost out of its patented period. It is a homogeneous assay (i.e. there are no washing or seperation steps involved). It works for small molecule analytes.

- EMIT. This is another small-molecule, homogeneous immunoassay, but one which is more sensitive than ARIS.

Other immunoassay formats fall into the **biosensor** category, which is considered to be more in the mainstream of biotechnology.

Immunosensors

Biosensors have a biological part and a detection part. The biological part confers selectivity on the sensor, the detection part detects the effect of whatever the biological part does and turns it into a recognizable 'signal' (usually an electrical signal). In immunosensors the biological part is an antibody. The physical part is usually a physical mass-detection system or an optical device.

There are two groups of immunosensors based on mass detection. Both use extremely small mass detectors, usually manufactured on a silicon 'chip' (and hence sometimes called 'microchip biosensors'), to detect the tiny changes in mass that occur when an antibody binds to an antigen. They are all resonance devices that measure the binding of the analyte to the probe.

The simplest type is based on the tuning form principle. The note that a tuning fork sounds depends on the mass of the tines. If the mass increase, the note goes down. The sensors have the equivalent of a microscopic tuning fork with the antibody coated on the tines. The silicon surface from which the tines are made detects the frequency with which they vibrate. When something binds to the antibody, the note falls and the circuit picks this up.

Surface acoustic wave (SAW) devices are a variation on this theme. Because the tuning fork is made of piezo-electric material, these are sometimes called piezo-electric sensors.

The problem with them is that anything landing on such a sensor gives a signal. Thus, despite having a very specific antibody as a biological element, they are very prone to interference. So while 'tuning fork'

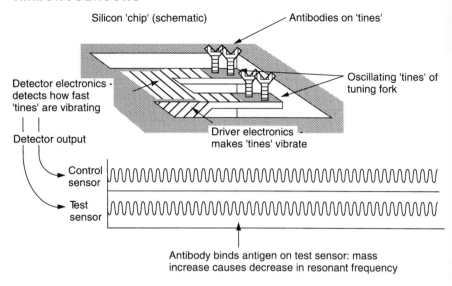

Silicon 'chip' (schematic)

Antibodies on 'tines'

Detector electronics - detects how fast 'tines' are vibrating

Oscillating 'tines' of tuning fork

Detector output

Driver electronics makes 'tines' vibrate

Control sensor

Test sensor

Antibody binds antigen on test sensor: mass increase causes decrease in resonant frequency

microdevices are known in mechanical applications such as strain gauges and gas sensors, they have not provided a reliable biosensor yet.

See also **Biosensor, Optical biosensor.**

Immunotherapeutics

These are drugs, usually biopharmaceuticals, which act on the immune system. Because the immune system regulates itself with a vast array of proteins that signal between the cells (the cytokines), most immuno-therapeutics are proteins made by genetic engineering to boost some aspect of the immune system, i.e. the way the white blood cells grow, differentiate, or act. Because the cells of the immune system produce only minute amounts of these proteins, to make them as drugs biotechnologists clone the corresponding genes. Many of them have only been discovered by cloning their genes and then seeing what the resultant protein does.

Among the ones being developed as drugs are:

- Interferon: second oldest biotechnological protein, being used as an immune system booster for many diseases.

- Interleukins: especially interleukin-2 (IL-2).

- CSFs (Colony stimulating factors). These stimulate the growth of the cells which make the white blood cells which are responsible for the immune response.

See also **Cytokines**.

Immunotherapy

This is the use of antibodies or antibody-derived proteins in treating disease. The use of antibodies as targeting agents (in immunoconjugates or immunotoxins, for example) is usually not considered to be immunotherapy. Rather, immunotherapy means giving the patient an antibody which they cannot make for themselves, because their immune system cannot work fast enough, because their immune system is not working at all because of disease, or because the antibody is against an antigen which the body would not normally recognize as 'foreign'.

For example, Xoma and Centocor have developed immunotherapeutic antibodies to treat sepsis (uncontrolled bacterial infection of the blood). The antibody binds to the endotoxin which the infecting bacteria produce, and which causes the symptoms of the disease. Sepsis develops within 24 h, far too short a time for the body to get its own immune response going, so the injected antibody fills the gap. The Centocor product received FDA approval for drug use in late 1991. (Celltech have attacked the same disease with an immunotherapeutic, but using a different target antigen. Their antibody is against tissue necrosis factor, which mediates some of the effects of the endotoxin).

Other immunotherapy targets are AIDS and meningitis.

Immunotherapy can also mean using whole cells from the immune system as a therapy. This latter has been tried under the title of adoptive immunotherapy, when NK cells ('natural killer' lymphocytes, some of the white cells in the blood able to destroy other cells) were taken from terminal cancer patients, stimulated using cytokines to become more energetic, and then injected back into the patient. The therapy had some effect, but severe side-effects. A variant of this is to use another class of white cells — tumour infiltrating lymphocytes (TILs) — which can target a cancer more specifically. Again, they must be obtained from the patient first.

TILs have also been tagged with foreign genes in a first approach to gene therapy in treating terminal cancer. The initial gene experiments put a 'useless' gene into the cells: the ultimate idea is to put a gene into TILs which will increase their effectiveness at killing tumours.

Immunotoxins

Immunotoxins are protein drugs. They consist of an antibody joined onto a toxin molecule. They have not been used as drugs on people yet, but show promise for treating some cancers in the future.

The toxins used — from bacteria diphtheria, pseudomonas or shigella or the castor bean toxin ricin — are extremely poisonous. Probably only a few molecules of ricin inside a cell can kill it. Thus they are of no use as systemic drugs. However, if they can be placed at a specific site, then they can be used to kill off one type of cell with very high efficiency. This is the idea behind immunotoxins. The toxin is joined to an antibody molecule which can bind specifically to one type of target cell. The resulting 'conjugate' is injected into the blood at extremely low concentration. When it encounters its target cell, the conjugate binds to it, concentrating the toxin there. The toxin then has a much higher chance of killing the cell. ImmunoGen has a rich-based immunotoxin of this sort in clinical trials as a treatment for leukaemia.

Refinements use parts of the toxin molecule, not all of it. Most toxins consist of a part which enables the toxin protein to enter the cell (the A chain) and a part which kills the cell (the B chain). Without either, the toxin is relatively ineffective, as the A chain is not toxic and the B chain needs to get inside the cell to work. Conjugating the B chain to an antibody makes a much less dangerous material: however it can still kill cells if the antibody binds to them, as the local concentration of B chains around that cell is so high that a few B chains get inside anyway by chance.

Immunotoxins have some limitations. As large molecules, they cannot get inside solid tumours easily. They are also rapidly mopped up by the patient's own immune system unless the patient is on immunosuppressive drugs. Also, there are some cells which bind antibodies non-specifically as part of the normal immune reaction. They will bind the immunotoxin, and so be killed.

Immunotoxins can be made by linking toxin and antibody molecule chemically. They can also be made by making a fusion of the genes for the toxin and the antibody: the resulting fusion protein is more stable, and can be smaller and less prone to binding to other tissues than a chemical conjugate. The antibody can also be 'humanized', reducing other complications.

A related idea is to use the toxins themselves as biotherapeutics (*see* **Toxins**).

In vivo vs in vitro

These latinisms are widely used when scientists are talking about doing something 'simple' in the laboratory and then taking the result and applying it to a more complicated, living system.

In vivo. Literally means 'in the living', and means in a living system, such as a complete animal. It is contrasted with *in vitro*, literally meaning 'in glass': this is translated by every English newspaper to 'in the test-tube'. It means 'in the laboratory', and is taken to mean the opposite of '*in vivo*'.

There is no clear ruling about whether cells are *in vivo* or *in vitro*: it depends on what you are talking about. The terms are usually used to contrast one experiment with another, rather than as absolute definitions.

Induction

In biotechnological terms this means getting an organism to make a protein, usually an enzyme, by exposing it to some stimulus. Usually the stimulus is a chemical, often a substrate for growth that is broken down by the induced enzyme. Induction involves the control of gene expression, but it is not a strictly genetic phenomenon, as no new genes or gene rearrangements are involved. It is only the expression of genes already there.

In general, an inducible gene, i.e. one that is capable of induction, can be induced by one or a few compounds. These are called inducers. These compounds (or sometimes their metabolites) affect how a protein binds to the promoter region of the gene concerned, and so affect the control of that gene. The exact mechanisms involved are as varied as they can be (as is the case in nearly all biology). Thus in order to be inducible, a gene needs to have the right promoter region. Some expression vectors have inducible promoters in them. They must also carry the genes for any proteins involved, of course — the inducer does not bind to naked DNA on its own.

A related term is repression. In repression a compound has the opposite effect to an inducer, reducing gene activity and so making the cell lose an enzyme activity. Such genes are called repressible. This can be very important in biotechnology, as many genes for useful enzymes such as those that make antibiotics and other secondary metabolites are repressed by common substances such as glucose.

Induction also means a form of logic that reasons from specific examples of something to general rules about that something. This is something biochemists do a lot, but that is rarely what they mean by induction. Despite the fact that it is logically indefensible, it seems to work.

Inoculation

Inoculation (apart from in the sense of immunizing someone) means introducing a small culture of a microorganism into a new environment with the intention that it should grow there. Thus fermenters are inoculated at the start of a run with a batch of organisms that have been grown to a state where they are ready to grow rapidly in the conditions provided by the fermenter. This may take some skill to achieve, as the conditions under which the inoculant was grown are probably not the same as those inside the final fermenter, and so the organisms could be adapted to a rather different culture condition.

The small dose of organisms (typically between 1 and 10 per cent of the number of organisms expected in the final fermentation) is called the inoculant.

The above refers to inoculating in the laboratory or production plant. Bacteria can also be inoculated into soil (to help bioremediation or to colonise the roots of plants), or onto plant roots or seeds directly. Again, the aim is to get them to grow in their new environment.

ISFET

Ion-sensitive field effect transistor. A field effect transistor (FET) is a semiconducting device in which the electric field over a 'junction' is used to modulate the current flowing through that junction. (The 'junction' is a region between different zones of the silicon crystal, usually ones with different impurities introduced into them, and which has a high electrical resistance unless an external electric field modifies its electronic properties.) It is a standard component of integrated circuits. Closely related in terms of its electronic effect is the MOSFET — metal oxide semiconductor FET.

It may be made into a sensor device by allowing ions to accumulate above the junction region. If the material above this region absorbs some ions specifically, then they will accumulate there and build up a charge. This will create an electric field, and so the FET will 'switch on'. A current will flow. Thus this device — an ion-sensitive FET, will allow a current to flow that is dependent on the amount of a specific ion present.

These devices may be useful as a method for monitoring ion concentration in a range of biotechnological processes. However they have also been turned into biosensors by replacing the ion-selective layer with an enzyme which generates ions by its action. Urease is a common example,

ISFET

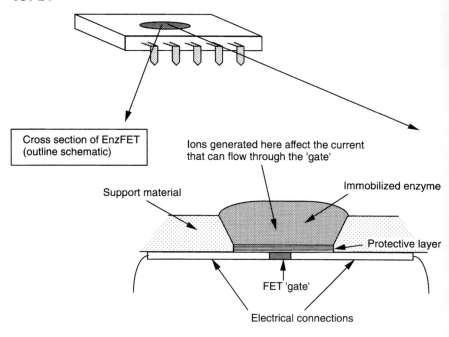

Cross section of EnzFET (outline schematic)

Ions generated here affect the current that can flow through the 'gate'

Support material

Immobilized enzyme

Protective layer

FET 'gate'

Electrical connections

as it takes uncharged urea molecules and splits them into ammonia and carbon dioxide: the ammonia picks up a proton to become charged ammonium ions, which the electrode detects. This sort of device is also called an enzyme FET (EnzFET or ENFET).

The attraction of ENFETs is that they could be manufactured by the large-scale manufacturing processes used by the semiconductor industry. The drawback is that they are too unreliable and too difficult to manufacture to be usable in most cases. A few exceptions use a FET as a detector for urease, the enzyme being used as a 'tag' to track the presence of some other molecule such as DNA or an antibody.

Claimed advantages of ISFET-based sensors include:

- They can be mass produced by the techniques of silicon-chip manufacture.
- Several sensors can be put on one chip, together with control and reference electrodes.
- The extremely small size of the device means that it can measure extremely small charge changes, and consequently is very sensitive.

While all these are true of the semiconductor base of the sensor, they have not yet been proven true of the whole device except in some research laboratories.

See also **Biosensor**.

Langmuir–Blodgett films

These are thin films of molecules formed on the surface of water. The original Langmuir–Blodgett film was a lipid layer on top of the water, but the term is often used to describe lipid films in which both sides are in water, or these films when they are transferred to a solid surface.

Lipids have a polar, water-loving (hydrophilic or lipophobic) 'head' and a water-hating (hydrophobic or lipophilic) 'tail' (*see* **Hydrophobicity**). Thus half of the molecule is soluble in water, half is not. The most stable arrangement of such molecules is to have them arranged in clusters with the tails on the inside away from the water and the head on the outside. One such cluster arrangement is a flat sheet, with the tails in the middle and the heads on either side. This is the Langmuir–Blodgett film, or lipid bilayer. It is the basis of the membranes that surround living cells and some of the organelles inside cells.

Lipid bilayer films or membranes are only one example of 'liquid membranes', in which a thin layer of liquid is stabilized so that it can last for a long time in water. They all have to be stabilized by some chemical means, as otherwise they collapse into little globules of liquid, or dissolve in the water.

Lipid bilayer membranes have applications in **drug delivery** systems (as liposomes), in biosensors, in separation processes and in some bioreactors. Nearly all of these applications are still only laboratory demonstrations.

Biosensor applications rely on the high electrical resistance of a Langmuir–Blodgett film, or on its optical properties.

Electrical sensors are based on the ability of some proteins to carry ions across a lipid membrane. Some antibiotics, proteins from nerve cell membranes, a variety of 'transport' proteins which allow cells to get materials from outside the cell into the cell without making holes in the membrane, can all be inserted into the membrane. The protein can allow one material or type of material — an amino acid, a metal ion, or maybe simply protons — to cross the membrane: in the presence of that material the membrane will conduct electricity. In its absence, the membrane will have a much higher resistance, because there will be no path for any other charged species to cross it. Thus the membrane could be a very sensitive detection system.

The problem with this is that the membranes are mechanically and

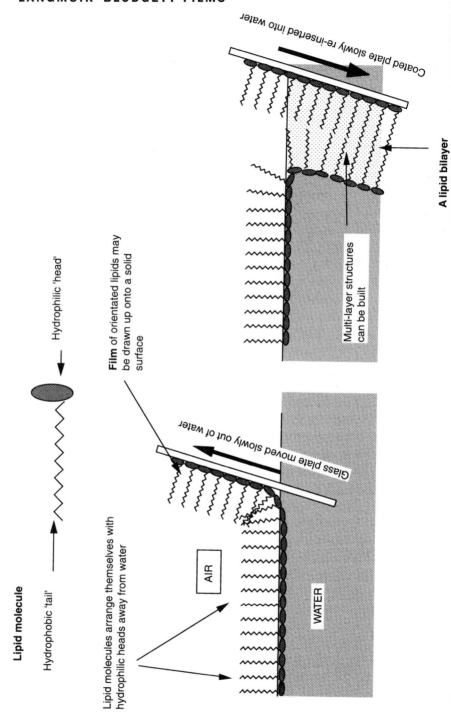

Lipid molecule

Hydrophobic 'tail'

Hydrophilic 'head'

Lipid molecules arrange themselves with hydrophilic heads away from water

AIR

WATER

Film of orientated lipids may be drawn up onto a solid surface

Glass plate moved slowly out of water

Coated plate slowly re-inserted into water

Multi-layer structures can be built

A lipid bilayer

chemically unstable, as are most of the proteins that we want to put into them. Thus, while a sensor system may be 'lashed up' in the laboratory, none have worked 'in the field'.

A related use for Langmuir–Blodgett films is as switching elements in computer-like circuits.

An alternative sensor system based on Langmuir–Blodgett films is an optical sensor system. Because the films are extremely thin, they cause interference effects when light is shone through them or reflected off them, and these effects depend critically on how thick the film is. If antibodies are immobilized on the surface of the film, then when they bind to their antigen the total thickness of the assembly will change (from being film + antibody to film + antibody + antigen), and so the colour of the reflected light will change. Again, this can be demonstrated for some very simple, model systems in the laboratory, but not for any realistic sensor application.

See also **Liposome, Liquid membrane, Molecular Computing**.

Leaching

Microbial leaching, or bioleaching, is the use of microorganisms, usually bacteria, to isolate metals from mineral ores by solubilizing them and allowing them to be washed ('leached') out of the ore. This is often called bioleaching. Thus it is a method of mining, and is a major component of microbial mining (**biohydrometallurgy**) technology.

Many ores cannot be processed economically because the concentration of metal in them is too low. Some of these ores are low-grade ores which are discarded as waste during mining operations aimed at higher-grade ores. (The 'grade' of an ore depends mainly on how much metal there is in it, but also on how accessible that metal is. Clay has a very high content of aluminium, but it is extremely expensive to extract aluminium from clay.) However if the metal can be released as a soluble salt, then it can be washed out and collected without the ore having to be mined, crushed and smelted, as in a normal mining operation.

LEACHING

Leaching is also used to extract gold and uranium from normal ores (*see* **Gold and uranium extraction**).

Leaching can be achieved in three physical arrangements. Slope or dump leaching is where a pile of the metal ore on the side of a hill is sprayed with a bacterial culture at the top and the eluant with the metal is collected at the bottom. Heap leaching is similar, but the material is in an isolated heap, which is more common in most mining sites. *In situ* leaching pumps the bacterial culture into the centre of the ore body along pipes or tunnels and then allows it to filter down to the base, where it is collected.

Leaching is a chemical process. In some cases bacteria oxidize sulphur in the mineral to sulphuric acid, deriving metabolic energy. The sulphuric acid solubilizes the metal (copper sulphate is soluble, while the sulphide is not, for example), and so the metals wash out in an acid solution. In others the bacteria act directly on the metal ion, for example oxidizing uranium (IV) (insoluble) to uranium (VI) (soluble). The ore to be leached is sprayed with the bacteria in a suitable nutrient mix, which supplies all the other chemicals it needs for growth. Thus the bacterium is limited by the energy it can obtain from digesting the mineral, and thus digests the ore as fast as it can. Optimizing the nutrient mix is a critical factor in making a bioleaching process work at a commercially useful rate.

Lipases

Lipases are enzymes which break down lipids into their component fatty acid and 'head group' moieties. The lipases used in biotechnology are almost invariably digestive lipases, meant to break down the fats in food. However they can be turned to a number of different uses.

They can be used to break down complex fats into their components, which are then used to make other materials. This, however, is a relatively prosaic use.

More talked about is transesterification, a process where a lipase is used to swop the fatty acid chains between lipids without ever releasing significant amounts of fatty acid. This is a useful thing to do, as it enables

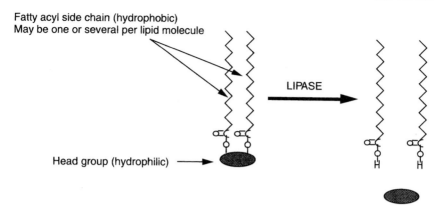

Fatty acyl side chain (hydrophobic)
May be one or several per lipid molecule

LIPASE

Head group (hydrophilic)

the biotechnologist to take a saturated fat (with a high melting point) and an unsaturated one (with a low melting point) and produce mixed molecules with intermediate properties: depending on how the ingredients are mixed, the properties can be determined fairly accurately. This requires that the lipase works in organic solvents, otherwise the enzyme would simply chop the lipids up.

Transesterification of triacylglycerol fats (the normal 'fat' in animal tissue) which is specific to the 1 and 3 fatty acids is a relatively specific and widely used transesterification, and is called interesterification.

See also **Organic phase catalysis**.

Liposome

A liposome is a small capsule made of lipids. Lipids can form stable sheets of molecules in solution, in which the polar 'heads' point outwards into the watery solution and the apolar 'tails' stick together in the middle of the sheet — this is a Langmuir–Blodgett film (*see* **Langmuir–Blodgett films**). If such a film closes up into a ball, the result is a sphere with watery solution outside and inside separated by a lipid 'bilayer'. This is a liposome. Liposomes can have multiple layers stacked inside each other, but are often considered as if they were single (unilammellar) bags.

Liposomes have been suggested as the basis of several methods of drug delivery, especially for delivering peptide drugs. This is because they could protect their contents from digestion in the stomach and so deliver them to the intestine, where they would be absorbed, or could allow them to be injected into the bloodstream and be carried around to a specific organ. Here the organ would recognize the lipids and absorb them specifically (this would work well for the liver, which tends to absorb liposomes from the blood spontaneously). Alternatively antibodies linked onto the outside of the liposome could bind it to the relevant tissue. Liposomes tend to accumulate at sites of inflammation, and in some tumours (no one knows why), and so they are potential delivery vehicles for anti-inflammatory and antitumour drugs.

Liposomes are considered particularly useful for this sort of application as they are made of the same materials (lipids) as the outside of cells, and so are less 'alien' to the body.

Trapping things inside liposomes is a form of encapsulation, and as such can be used in many other areas. Here, however, liposomes are less

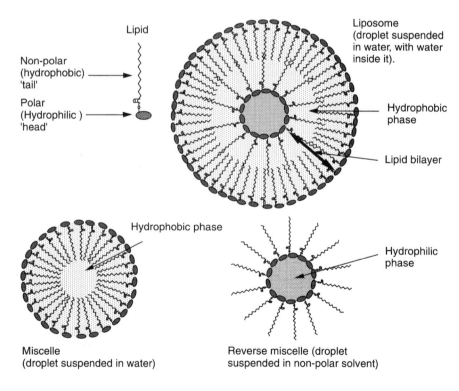

favoured because they can be rather less stable than other, polymer-based encapsulation methods.

Liquid membrane separations

Liquid membranes are thin layers of a liquid which does not mix with the water on either side. (In principle, they could also be thin layers of water with some other liquid on either side, too.) If something can dissolve in the liquid, then it can pass through the membrane. This can be the basis for separating materials which do dissolve in the liquid from those that do not, by putting the mixture on one side of the membrane and pure water on the other: the soluble component diffuses across the membrane, the contaminants cannot.

More sophisticated separation mechanisms can be based around this idea. The membrane can be impregnated with a carrier molecule which can carry across one type of molecule but not others. These usually bind to the target molecule, making it soluble in the lipid (as a 'complex') whereas it usually would not be. Chemicals which can do this could include some peptide antibiotics, clathrins, crown ethers, or cyclodextrins. The transport of the molecule you want can also be linked to the transport of another molecule (for example, a proton): this is called 'coupled transport', and is the way in which living cells concentrate many molecules inside themselves.

Ion exchange systems can also be used with a supported liquid membrane, in an ion exchange membrane (IEM).

Liquid membranes

These are thin films make up of liquids (as opposed to solids) which are stable in another liquid (usually water). Thus the liquid must not dissolve in the water, but nevertheless must be prevented from collapsing into a lot of small droplets. There are several types of liquid membrane.

- Langmuir–Blodgett films. These are 'true' liquid membranes, in that there need be nothing in them except the liquid (see Langmuir–Blodgett films).

- Immobilized or supported membranes (immobilized liquid membranes — ILM). Here the liquid is trapped in a thin film of some solid material. This might be a porous polymer (such as scintered glass) or a fibrous one (such as cellulose). The liquid fills up the pores in the material, and so forms a series of mini-membranes.

- Supported membranes can also be ion-exchange membranes (IEMs), if the supporting material is something which binds ions strongly. When something dissolves in the liquid part of the membrane, it is held there by the solid part. This can be the basis of separation methods.

- Emulsion liquid membranes (ELMs). Here the watery part and the non-watery liquid are mixed up with a detergent. This makes small droplets of water in the other liquid (or the other liquid in water) stable. The result is a mix of water inside liquid droplets, themselves inside water. This is a membrane, as it is a liquid barrier between two volumes of water.

Liquid membranes can be used in a number of applications. Their main potential is as the basis of separation systems (*see* **Liquid membrane separations**).

See also **Langmuir–Blodgett films**.

Live vaccines

Live vaccines are vaccines containing living organisms or intact viruses, rather than inactivated (killed) organisms or extracts of them. They can cause better immunity in patients, but have the potential drawback that, unless they are thoroughly 'crippled' in some way, they may cause disease.

Biotechnologists have generated ideas and research products for live vaccine development in a number of areas. Viral vaccines are dealt with elsewhere (*see* **Viral vaccines**). Bacterial live vaccines can be developed in a number of ways.

Attenuation. Bacteria need a number of specific genes (virulence genes) to be able to cause disease, but these genes are not essential for growth in the 'test-tube'. When pathogenic bacteria are grown outside their host, they tend to loose their virulence genes by mutation. The result is an attenuated bacterium which will cause an immune response similar to the original but which is harmless. Usually several mutations are needed to make sure that a strain is really attenuated. If the nature of the virulence genes is known, then conventional and 'molecular' genetics can be used to select for mutations in, or loss of, those genes.

Gene cloning. An alternative is to place some key genes from the pathogenic bacterium into another, harmless organism. These may be the genes for surface parts of the pathogenic bacterium such as pili or transport proteins which are seen by the immune system. The degree to which an antigen, or a particular part (epitope) of an antigen is detected by the immune system, and hence the amount of antibody response which the immune system makes against that antigen, is called its immunogenicity. A key part of the design of a good vaccine is deciding how to make the vaccine highly immunogenic, so that it is 'seen' by the immune system very clearly.

On vaccination with such a material, the immune system 'learns' what the clonal molecules from the pathogen look like without ever encountering the whole organism. This is similar to cloning the protein as a vaccine, but has the advantage that, as part of a living organism, it can stimulate the immune systems to greater prodigies of invention in generating good antibodies against it.

Live bacterial vaccines are generally discussed for fighting enteric infections, including tooth decay, and some parasitic diseases.

Loop bioreactors

Also loop fermenters, these are bioreactors in which the fermenting material is cycled between a bulk tank and a smaller tank or loop of pipes. The circulation helps to mix the materials and to ensure that gas injected into the fermenter (usually oxygen or air) is well distributed among the liquid. The reactors are also very useful for photosynthetic fermentations, where they allow for the photosynthesizing organism to be passed along a large number of small pipes, where the light can get to them easily, rather than inside a single volume, where only the organisms near the edges get much light.

There are lots of types of loop bioreactor, but they break down into those which have an internal loop (e.g. a stirred tank bioreactor with an internal draft tube) and those which have an external loop. Some airlift fermenters are of the first type, as are pressure cycle reactors — reactors where air or oxygen is injected into the riser half of the reactor and this drives the liquid in that half up, so pushing the flow round the vessel. A variant that is found in all of them is the jet loop reactor, in which the recycled liquid is injected with considerable force back into the main tank. This means that it does not merely circulate the re-injected liquid around, it stirs the rest of the tank contents up as well. This has the advantage that the recycling mechanism is also the stirring system, removing the need for stirrers and baffle plates.

One popular type of loop bioreactor is the airlift bioreactor (or fermenter).

See also **Airlift fermenter**.

Luminescence

Luminescence, the production of light by chemicals, is gaining increasing use as a labelling system for antibody- and DNA-based tests. Luminescent tests are of interest because, if carried out in a rigorously light-proof box, they can be extremely sensitive: a photomultiplier tube can detect when only a handful of photons have been given out by a reaction, so offering the potential for detecting only a handful of DNA or antibody molecules.

There are two broad ways of generating light using chemicals:

- Chemiluminescence. This uses specific chemical groups that, when reacted, give out light. They can be attached to many other chemicals (e.g. proteins, DNA). There are also chemiluminescent groups which have phosphate groups attached to them. As they are, they cannot react to give off light. However when the phosphate group is chopped off, they become potentially chemiluminescent. This allows a chemiluminescent reaction to be used to detect an enzyme that cleaves phosphate groups, such as the widely used alkaline phosphatase (AP). AP is often used as a reporter group for enzyme immunoassays (EIA). Adding chemiluminescence to such an assay enhances its sensitivity greatly.

- Bioluminescence. Some specialized enzyme systems can also generate light, using the energy of ATP (adenosine triphosphate) to do so. These enzymes are called luciferases. The most commonly used luciferase is derived from bacteria. Enzymes from fire flies have also been used.

Maxicells

Maxicells are bacterial cells that have a mutation in the genes that control how they divide. Under the 'right' conditions (usually when the temperature of the medium is raised) they simply stop dividing. However they do not stop growing, so the result is a huge bacterial cell. This can be helpful, in that such large cells are much easier to separate from their medium than the relatively small normal cells: they will, for example, settle out of solution under their own weight in a fairly short time.

A related idea is the minicell. This is another cell division mutant. Here, under the 'right' conditions the cells divide, but they do not divide in the middle. Rather a small 'minicell' splits off one end. All the bacterial DNA stays in the main cell, so the minicell has no DNA of its own. Therefore it cannot make any new RNA, and as soon as the RNA that happened to be in the cell has broken down it cannot make any new protein either. However this rule is broken if the cell contains certain types of plasmids — a few plasmid molecules *can* end up in the minicell. Thus, when all the trapped RNA has broken down, the only proteins being made by the minicell are those coded by genes on the plasmid. This is very useful for studies of gene expression, as, by isolating minicells, the proteins being made 'by' the plasmid can be studied without having to purify them from all the other proteins being made by a normal bacterial cell.

Microbial mining

This is the use of microorganisms to remove minerals, and particularly metals, from rocks. It is a specific application of **biohydrometallurgy**. Microbial mining is related to the use of microbes in desulphurization and for bioremediation (*see* **Desulphurization, Bioremediation**).

Microbial mining falls into two areas:

- Leaching. This is the use of bacteria to process ores to make the metals in them more accessible. Usually it involves using bacteria to release the metals as soluble salts, to be washed out for subsequent recovery. However it can also involve pre-processing of the ores which, while not releasing metals directly, allows them to be separated more easily by washing out, flotation or another 'traditional' method in a further processing step (*see* **Leaching**).

- Purification. Using microorganisms or microorganism components to separate and concentrate metals from very dilute solutions. This is also called biosorption (*see* **Biosorption**).

Biohydrometallurgy is used commercially to recover copper and uranium from low-grade ores, especially chalcopyrite ($CuFeS_2$), covellite (CuS), chalcocite (Cu_2S), and uraninite (UO_2). A number of other metals (antimony, arsenic, molybdenum, zinc, cadmium, cobalt, nickel, and gold) can be extracted using bacteria, but these are not used to a significant extent. Bacteria of the groups *Ferrobacillus* and *Thiobacillus* are commonly used in extraction processes involving the oxidation of sulfides.

Microbial processes are also used to recover oil, either by altering the oil's properties underground (notably altering the pH), or by generating 'mud' underground. This is a general name for viscous solutions which are pumped into a well to force oil to the surface. The problem is that considerable pumping is needed to get the stuff down the well in the first place. Microbial mining systems, aimed at pumping a highly fluid bacterial mix down the shaft which then synthesizes extracellular polymers to create a thick solution underground, sound plausible but have yet to be demonstrated in realistic trials.

Microcarriers

In biotechnology, microcarriers are generally small particles used as a support material for cells, and particularly mammalian cells, in a large-scale culture. Mammalian cells are too fragile to be pumped and stirred as bacterial cells are, but still need feeding with oxygen and nutrients and must still be separated from their culture medium when the time comes to collect the product.

In mammalian cell culture, microcarriers are particularly useful for culturing cells which would normally grow attached to a solid surface (attached or surface culture, as opposed to suspension culture). Rather than have to have acres of flat plastic surface, the cells are grown over the surface of small polymer spheres of plastic, especially, polystyrene, gelatin, collagen, or polysaccharides like dextran or cellulose. The surface area for growth is huge, and the spheres can be treated like bacterial cells for filtering and for (gentle) centrifugation, and protect the cells from the shear forces involved in pumping and aeration. Some microcarriers are simply solid spheres, some are porous. Porous ones have a larger surface area for cells to grown on, and the cells can grow inside them as well as on the outside, so giving them greater protection. However it is harder to see the cells on the carriers, which can be important if you want to know how well your culture is growing.

An alternative to growing cells on microcarriers is to grow the cells as aggregates. Cell aggregates have some of the mechanical robustness of cells on microcarriers, but have much higher cell content for a given amount of solid matter. However getting cells to grow in aggregates can be much harder than getting them to grow on suitably treated polymer surfaces.

Microorganism safety classification

One major concern about biotechnology is whether it is safe. As much of biotechnology involves the genetic manipulation, selection, or physiological manipulation of microorganisms and their subsequent production in large amounts, some of this concern translates into a concern about the safety of industrial-scale microbiology.

Most guidelines and codes of practice in handling microorganisms are aimed at medical microbiologists and microbiologists handling pathogens to produce vaccines. Thus many of the guidelines as to how microorganisms should be handled in biotechnology are derived from these medical examples. The World Health Organization has no evidence that genetically manipulated organisms are inherently more dangerous than any others, and has found no case where a laboratory or industrial worker has contracted an infection as a result of contact with a genetically engineered organism.

The principle of classifying the danger from a microorganism, and so of deciding how to contain that danger, is to classify the organism according to how likely it is to escape, how likely it is to survive if it does escape, and how much damage it could do if it does survive. Different countries have different rules about how this is done: the table below summarizes a few of these.

Institution	Risk: Minimal	Normal microbiological risks	High risk to individual only	High risk to individual and to community
ACDP*, ACGM‡¶	—Group 1—	—Group 2—	—Group 3—	—Group 4—
EFB+	Class 1	Class 2	Class 3	Class 4
WHO	Group I	Group II	Group III	Group IV

*Advisory Committee on Dangerous Pathogens (UK). +European Federation for Biotechnology, which has the same groupings as the US Public Health Service (PHS). ‡Advisory Committee on Genetic Modification (UK).

If an organism is outside the Class 1/Group 1 area, then it can be contained by a variety of physical or biological methods.

A range of national safety committees oversee that appropriate containment is used for biotechnological applications of organisms in each class (even if, in other industries, no containment is needed for those same organisms at all).

See also **Biological containment, Clean room, Physical containment**.

Microorganisms

There is a very wide range of microorganisms used in biotechnology. *E. coli* and the yeast *Saccharomyces cereviseae* are mentioned in many places in this book. However a range of other organisms is frequently used in biotechnology.

Microorganisms, indeed all life, is classified into prokaryotes (organisms without a cell nucleus) and eukaryotes (organisms with a cell nucleus). Animals, plants, and fungi are all eukaryotes; bacteria and archebacteria are prokaryotes.

Bacteria are classified into Gram-positive and Gram-negative. These names reflect whether their cell walls will absorb Gram's stain, but the division they represent is quite a substantial one, and Gram-positive and Gram-negative organisms are biochemically and genetically quite different. However they may *look* much the same under the microscope. Microorganisms may be ball-shaped (cocci), rod-shaped, or made of very long 'strings' called hyphae. Hyphae may be branched or unbranched: in either case they are often harder to grow in bulk because the stirring needed to get nutrient to all the hyphae can break them. Organisms that grow as long strings or filaments are called 'filamentous'.

Microorganisms are also classified into aerobes (grown in the presence of oxygen) and anaerobes (which do not use oxygen). These can be facultative and obligate: facultative aerobes can use oxygen or not: obligate aerobes must have it. Obligate anaerobes are killed by oxygen.

Some of the more commonly mentioned organisms are:

- *Aspergillus.* A filamentous fungus that has been used for genetic engineering in a few cases, and which is also used to produce citric acid by fermentation.

- *Bacillus subtilis.* This Gram-positive bacterium is also widely used as a cloning host, especially for the expression of secreted proteins. Strains that lack any protease activity have been developed, which therefore do not break down the product protein when it is secreted into the fermentation medium.

- *Candida utilis.* A yeast, this organism is used in fermentations to produce chemicals.

- *Clostridium acetobutylicum.* A bacterium used in the past to produce acetone and butanol by fermentation, and now used as a source of enzymes.

- *Escherichia coli.* Usually abbreviated to *E. coli* in print (and almost always in conversation), this very versatile Gram-negative bacterium is used in many biotechnological processes. Its genetics are the best known of any organism, with the majority of its genes known and about 30 per cent sequenced. It is by far the most common host cell for recombinant DNA work. It is also used in fermentations to make many amino acids and other products, as it grows on many, very cheap fermentation substrates, grows fast, and can be manipulated genetically to accumulate many different chemicals. It is also very chemically versatile and quite non-pathogenic (with the exception of a few strains that, obviously, are not used for biotechnology).

- *Penicillium.* A group of filamentous fungi used primarily to produce penicillin antibiotics.

- *Pseudomonas.* A group of soil bacteria that contain some extremely diverse chemical abilities, that biotechnologists have harnessed in **bioremediation**.

- *Saccharomyces.* A group of yeasts. *Saccharomyces cereviseae* is brewers' and bakers' yeast, and as such is probably the most widely exploited microorganism. *Saccharomyces* are also used in recombinant DNA work as they are eukaryotes, and hence have the same sort of genetic structure as humans, secrete proteins in a similar way and so on, but are almost as easy to ferment in bulk as bacteria.

- *Streptomyces.* Gram-positive bacteria that are used to produce a range of chemicals, especially antibiotics. They have also been used as the hosts for genetic engineering, in part to manipulate their antibiotic synthetic pathways.

Also mentioned elsewhere in the book are Agrobacterium tumefaciens, Thiobacillus and Ferrobacillus (microbial mining), Methanococcus (single cell protein).

Micropropagation

This is a term used in plant production for the use of biotechnological methods to grow large numbers of plants from very small pieces of plants, often from single cells using tissue culture methods. In essence, a desirable plant is cut up into many very small pieces (sometime single cells, sometimes clusters made of a few to thousands of cells), and cultured. The culture conditions are tuned so that the cells grow into a callus, a mass of cells which look like a small mould. The conditions are then switched so that the callus starts to develop into a small plant 'embryo' (*see* **Embryogenesis**). Once this embryo has grown sufficiently, it can be planted out as a small plant. In some techniques, the embryo is encapsulated in a protective sheath so that when it is sown it has a 'shell' similar to the seed produced by more conventional breeding.

The advantages of micropropagation are that it can produce a very large number of plants in a relatively small space of time, and that all the plants are usually genetically identical. Among the drawbacks are that it is skill-intensive, and hence much more expensive than conventional breeding, and that it can only be done on plants for which the right cell culture conditions have been worked out.

The major drawback, however, is that during the callus phase the plant tissue can undergo enormous genetic rearrangement, often doubling up its chromosome number, or loosing bits of chromosomes or even whole chromosomes. This gives rise to the phenomenon of somaclonal variation.

See also **Somaclonal variation**.

Molecular biology

Much of biotechnology is based at least in part on molecular biology. But what is molecular biology?

Molecular biology, and its twin science molecular genetics, started in the late 1940s around a group of biologists and physicists-turned-biologists who were looking for a new way to attack the fundamental problems of life. Traditional biochemists at that time (and many biochemists today) sought to attack complex systems by taking them apart and analysing all the bits as carefully as possible in terms of chemicals and biochemicals. The 'new' scientists instead chose the simplest system that they could and sought to analyse it using genetics as their primary tool. The system they choose was the bacteriophage, and hence many of the founders of molecular genetics were members of a semi-official 'phage group'.

The genetic approach paid off handsomely in three ways.

- Firstly, it opened up whole new areas of genetics — genetics at a molecular level rather than the whole organism genetics that had characterized previous work on drosophila, plants and so on, or the biochemical genetics of bacteria and yeast. This in turn allowed the investigators to start to decipher the genetic code, deduce some of the mechanisms of protein synthesis, etc.

- Secondly, and more importantly, it gave credibility to a new way of thinking of biology. This way is now entrenched in the normal way of thought, and envisages the molecular basis of biology as being made up of quite understandable building blocks that bump into each other and latch onto and off each other in defined ways. Whereas in 1950 an enzyme was a squiggle in an equation, in 1990 it is a coloured 'blob' on a computer graphics display. The molecules at the basis of life became much more real and much more important. Life became a discrete machine, and the instructions for that machine lay in DNA, Hence the centrality of DNA to much of biology today. This approach to living systems, as discrete blocks labelled 'protein' and 'gene', has been called 'molecular Lego'.

- Thirdly, the phage group work gave us the basic tools of recombinant DNA technology. Thus, restriction enzymes, DNA ligase, and many cloning vectors all come directly from bacteriophage genetics.

Thus molecular biology is not a science in the sense that it studies molecules, or biology — biochemistry, physiology, pathology, microbiology do that too. It is more a way of doing biology, both a way of thinking about it and of getting the tools to do experiments. It is, in Thomas Kuhn's term, a paradigm. (It may also be a wrong paradigm — computer scientists, after thinking that intelligence was like Lego or a computer program for forty years, are now moving towards thinking that it is nothing of the sort.)

The combination of the ability to manipulate DNA as a common chemical and to think about the result in terms of computer programming or Lego construction has built much modern biology, and hence much biotechnology.

Molecular computing

An avant-garde area of molecular sciences which has involved some biotechnological thought, this term means making electronic or computing devices out of single molecules, or small groups of molecules. Talk about switches which are made of a single protein molecule leads to computers with greater-than-human powers which could fit in a matchbox. Fairly obviously this is speculative, but may not be as speculative as it sounds.

Firstly, proteins have already been used to build nanometer-scale patterns on microchip surfaces in research. These are not functional chips, but show that proteins could be used to assist building more conventional semiconductor devices, because they can self-assemble complex arrays of molecules on a surface which can then be used as the basis for deriving the chip's electronic properties. In early 1992 a protein layer on an electrode was shown to act as a diode, a critical if simple element of a logic circuit.

Secondly, many proteins do have charge-transfer and charge-switching properties, which could, with much greater understanding of the properties of proteins in general, be harnessed to provide some aspects of the information processing capability of a semiconductor device.

Thirdly, Langmuir–Blodgett films — thin films of lipids — are known to be an essential part of the electrical properties of nerve cells, and can be made quite readily in the laboratory. Nerve-cell proteins are inserted into the lipid film which alter the film's ability to let ions pass depending on what other ions are present or on the electric field to which they are exposed. This has progressed to the stage of building films, putting proteins into them, and demonstrating the electrical characteristics of the protein, which is similar to the position with transistors in 1930.

Molecular computing was a trendy term a few years ago, but has now been partly supplanted by nanotechnology. This is a related term, but means molecular-scale engineering rather than electronics. The most quoted idea is that of a tiny submarine which can be injected into a patient to clear out arteries clogged by atherosclerosis. Biologists could provide some of the elements of this (for example, the world's smallest screw propeller is the flagellum of a bacterium). However this is definitely twenty-first century stuff. Micromechanics, building engineering structures on silicon chips, works on a scale of tens of micrometres rather than the hundred-nanometre scale needed for nanotechnology, and has focused on a few well-defined products such as pressure and strain gauges. The success of micromechanics in a few fields does not imply that molecular electronics or nanotechnology are realistic in the next few years.

Molecular graphics

This is the display of molecular shapes, usually on computer. It has gained a lot of publicity because of its application to rational drug design. Molecular graphics takes the description of how the atoms of a molecule are arranged in space from a database and draws a picture of what the molecule would look like, for example if it were made of solid balls (the atoms) or thin sticks (the bonds between atoms). Usually molecular graphics does not calculate the structure of the compound.

Because the human brain is extremely good at perceiving patterns in complex pictures, but rather poor at seeing patterns in large collections of numbers, molecular graphics is a good way of allowing people to see the similarities in structures between molecules, and also of seeing whether two molecules fit together well. This is turn is useful when, as part of a rational drug design programme, a scientist wishes to find a molecule which will fit into the known structure of the active site of an enzyme, or the hormone-binding site of a receptor.

Molecular graphics packages often produce extremely attractive pictures as part of their output, which is another reason whey they have a high profile in the public relations material of pharmaceutical and biotechnology companies. The more sophisticated display methods can produce three-dimensional pictures which the user can manipulate as if he were in a room-full of bits of molecule which he can move around with his hands, a type of computer interaction called 'virtual reality'.

See also **Computational chemistry, Rational drug design**.

Molecular modelling

This is the use of computers to model what molecules look like. At one end of this range of techniques is molecular graphics, which is simply drawing three-dimensional drawings of what a molecule would look like, for example if the atoms were solid balls. At the other end it shades into computational chemistry — the calculation of what the physical and chemical properties of a molecule are. Usually molecular modelling falls more into the graphics end of the spectrum.

Using molecular modelling, rational drug design programmes can look at a range of different molecular structures of drugs which may fit into the active site of an enzyme, and by moving them on the computer screen decide which actually fit the site well. Molecular modelling can add sophistication to picture drawing by calculating the hydration (degree to which individual bits of the molecule bond neighbouring water molecules) and charge distribution across the molecule. These also affect how molecules bind together.

Monoclonal antibodies

Antibodies produced in the blood are made from a large number of different lymphocytes (B cells). Each B cell makes a unique antibody, so the antibodies which recognize any particular antigen are a mixture of molecules. This mixture is called 'a' polyclonal antibody: an antibody preparation that reacts with only one antigen, but which nevertheless derives from many different 'clones' of B cells. While this is useful for the body (covering the antigen with antibodies) it is a problem for the biotechnologist who wants defined materials to work with. Monoclonal antibodies are a way around this. These are antibodies made from a single clone of B cells which has been isolated and immortalized for growth *in vitro*. The invention of the methods for producing monoclonal antibodies won César Milstein a Nobel Prize. Milstein (and the Medical Research Council, who funded his work) did not patent the procedure for making monoclonal antibodies.

Monoclonal antibodies are generated as follows:

- Immunization — a mouse (usually) is immunized with the target antigen. This is done by injecting the antigen, sometimes with another material (an adjuvant) to stimulate the immune response (*see* **Immunization**).

- Splenectomy — the spleen (a useful concentrated source of B cells) is removed from the mouse.

- Fusion — the lymophocytes are fused with an immortal cell line. This immortalizes them, i.e. they will grow forever in culture.

- Cloning — the fused cells are put at very low concentrations into the wells of a multiwell plate. On average each well will only have one cell in it, so on average each cell will be a clone, i.e. derived from a single cell. This ensures that you have a pure cell line. This cell line is termed a hybridoma.

- Selection — the clones are screened by whatever method is to hand to find the one producing a good antibody against the antigen we want. Most of the cells will *not* be making the desired antibody.

A 'good' antibody is one which binds tightly to the antigen (in chemical terms, has an affinity of 10^9 or better), does not bind significantly to any-

MONOCLONAL ANTIBODIES

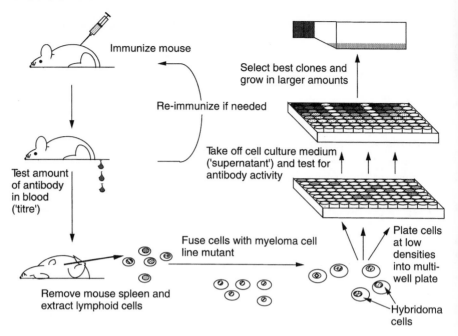

Immunize mouse

Select best clones and
grow in larger amounts

Re-immunize if needed

Test amount
of antibody
in blood
('titre')

Take off cell culture medium
('supernatant') and test for
antibody activity

Fuse cells with myeloma cell
line mutant

Plate cells
at low
densities
into multi-
well plate

Remove mouse spleen and
extract lymphoid cells

Hybridoma
cells

thing else, and is the right class and sub-class (IgG, IgM, etc.), although the exact choice of antibody will depend on for what purpose the scientist wants it.

If the target molecule is a very small one (like a drug molecule) then injecting it into a mouse will rarely produce an antibody response. In this case the molecule is chemically linked onto a larger molecule, usually a protein and often bovine serum albumin (BSA) or keyhole limpet haemocyanin (KLH), so that the immune system can 'see' it. The small molecule is called a 'hapten' in this case.

Most biotechnological applications use monoclonal antibodies unless they say that they use the 'natural' type obtained from the blood of immunized animals, which are called polyclonal antibodies.

See also **Antibodies, Binding**.

Monoclonal antibody production

Monoclonal antibodies can be produced commercially in a number of ways, depending on the scale of production.

As mouse ascites fluid. Mice can be injected with the hybridoma call line which makes the monoclonal antibody. Their ascitic fluid (the fluid which surrounds the lungs) or the blood plasma is collected and the antibody purified from this. This is simple, and does not require sterile culture conditions. However it does require animal facilities, and only produces around 50 mg/mouse. Thus it is used extensively for research volume production.

Tissue culture methods. Tissue culture methods which are used to make the hybridoma in the first place can be used to make antibody — the tissue culture 'supernatant', i.e. what is left of the medium once you have removed the cells, is a source of antibody. However this is rarely effective at producing more than 10 mg of antibody.

Suspended cell fermenters. The traditional biotechnology has been used to grow hybridoma cells in bulk. Celltech, for example, has a 1000 l airlift fermenter which can produce 100 g of antibody in a 2-week fermentation with a hybridoma. This is a similar technology to medium-scale microbial fermentation, although, because mammalian cells are very sensitive to chemicals, temperature changes, shear (squashing), and other environmental disruptions, it can be more difficult to get to work reliably, and also costs a lot more in expensive culture media.

Immobilized cell reactors. Several types of immobilized cell reactors have been used to make monoclonal antibodies on a scale of a few grams. The most popular is probably the **hollow fibre** reactor. A few grams of antibody is enough for several million tests to be used for medical diagnosis, for example, and so suffices most commercial needs.

Bacteria. An emerging technology involves the use of bacteria to produce antibodies. The genes for the heavy and the light chain must be spliced into one bacterium, but when this is done the bug is very much easier to grow than mammalian cells. This also makes genetic engineering of **chimeric** or **humanized antibodies** easier, as the cloning technology necessary to do this is done in *E. coli*.

See also **Antibody structure, Dabs, Monoclonal antibodies**.

Motifs

Neither protein nor DNA sequences are random. If nature 'wishes' to evolve a protein to do something, she starts out with proteins that already exist to do something else and, usually, shuffles bits of the relevant genes around to make the new entity. Thus certain strings of bases or amino acids crop up time and again in different genes and proteins. These are often called motifs. Usually they are significant because they denote that some bit of the molecule has a particular function. Thus 'zinc finger motifs' in proteins suggest that the protein has a section that binds to DNA. Similarly in DNA the 'TATAA' motif is suggestive of a promoter sequence in eukaryotic cells.

'Motifs' are similar to signal sequences in proteins. However signal sequences are meant to be 'read' by the cell. Motifs may have a functional significance, but may be of interest only because they give the biotechnologist clues to what a particular part of a gene of protein does. Among signal sequences known are the leader sequences that lead to secretion, another leader sequence that tags the protein as headed for the endoplasmic reticulum and lysosomes, the leader sequence that sends the protein to the cell nucleus, the 'stop transfer' sequence that anchors a protein in the cell membrane and so on. Being able to read signal sequences is also helpful, as it gives clues as to where in the cell a particular protein is meant to end up, and hence what its function might be. Signal sequences are only of relevance to proteins (although, of course, they are coded in the DNA), whereas sequence motifs can be found in DNA or protein.

Mutagenicity tests

There are a range of tests using biological systems to see whether compounds can cause mutations. It is argued that chemicals which cause mutations are also likely to cause cancer in man, a correlation which has generally found to be true. The main single cell test systems used are:

- Ames test. Named after Bruce Ames, this test exposed salmonella strains carrying specific genes to the chemical. New mutants are detected as bacteria which can grow without being provided with histidine ('black mutations'). The test is one of the standard battery of tests required for mutagenicity testing of products.

- SOS–chromotest. This is an alternative bacterial test which detects when an *E. coli* has had its DNA repair enzymes activated. Mutagens activate specific enzymes which repair damage in DNA, and the test uses a side-effect of these enzymes to detect their activity. Not generally accepted.

- Micronucleus test. This looks for a characteristic aberration of the chromosomes (formation of small fragments of genetic material outside the nucleus, which are called 'micronuclei') in cultured mammalian cells, usually chinese hamster ovary (CHO) cells.

Ames himself has recently said that most mutagenicity testing, including his own tests system, is largely irrelevant to human health, as 99 per cent of the mutagens and carcinogens we are exposed to come from the 'natural' environment and not from man-made sources.

Mythogenesis

Biotechnology has been extraordinarily successful in attracting scientists and investment. This has occurred despite that fact that few biotechnology companies are even today breaking even, and there are very few truly biotechnological products on the market that were not there ten years ago. One entirely reasonable explanation for this is that much of the new wave of biotechnology is aimed at medical problems, and these take a long time to solve, are great intellectual and social challenges, and could earn enormous profits for their conqueror. Another explanation is that this is a *post-hoc* rationalization for something much deeper, and that biotechnology is attractive because it promises the realization of very ancient dreams, in Jungian terms the physical embodiment of mythic archetypes.

Thus biotechnology has been taken to promise life extension via pharmaceuticals that are specific and 'natural' (both secondary metabolite products and biotherapeutics), the creation of plausible supermen, especially in sports applications, reproduction without sex, human cloning (and thus both a kind of immortality and the potential for children who are an extension of their parents), wild new animals such as chimeras and giants, and so on.

In literal terms this is nonsense — chimeric animals look like any other animals, 'giant' mice are only 30 per cent longer than normal mice, and human reproduction has never been so carefully legislated. However this need not matter. If biotechnology is perceived subconsciously as opening the doors to such archetypical dream worlds, then it will attract and repel much more strongly than if it is perceived just as a lot of scientists making money out of clever brewing. At a meeting in mid 1992 in the UK all the achievements of serious science were glossed over by the serious press in favour of reports of a scientist who claimed that he would be able to produce a cheese-flavoured cauliflower. (The non-serious press did not report the meeting at all, of course.) Why this focus, when it was only meant as a light-hearted example of what might be possible using plant genetic engineering? Because the 'allfood', the single food that is all you need to eat, has strong mythic roots going back to Greek ambrosia and Biblical manna, and anything that suggests that scientists are working on such an allfood is more attention-grabbing, even if it is nonsense, than the number of people dying of AIDS.

This could be important for the science and industry of biotechnology, as it suggests that much of the campaign waged to gain public acceptance of biotechnology may be based on false assumptions, and consequently will not convince many people. Indeed, it may even be counter-productive. By focusing public attention on mundane facts rather than mythic images, biotechnologists may reduce the public desire for biotechnology. A study on European attitudes to biotechnology carried out in 1991 may support this point, by showing that the more the people of a country were informed about biotechnology from the education that the government and industry put out, the more they were against it.

Names

One of the most fiercely competitive areas for biotechnology start-ups is finding a good name. As well as the obvious ones ('Monoclonal Antibodies Inc.', 'Affinity Chromatography Ltd.') biotechnology companies' names are assembled from a wide range of standard units. Start with one of:

- Bio-: almost an essential part. Means to do with life.
- Immu- or immuno-: to do with the immune system, usually to do with antibodies.
- Hyb- or hybri-: usually to do with DNA hybridization. Can also be related to making hybrid species. Hybritech is the maverick here, being concerned with monoclonal antibodies.
- Trans-: across, suggesting multidisciplinarity. 'Transgenics' is a special case.
- Eco-: now needs no introduction — anything that can be considered 'ecological'.
- Agro- or Agri-: agricultural.
- Myco-: to do with fungi.
- Onco-: to do with cancer.
- Cyto-: to do with cells (usually means mammalian cells).
- Gen-: to do with genes, and hence with recombinant DNA.
- Enz- or Enzo-: to do with enzymes.

and finish with one of:

- -gene or -gen: anything to do with genes.
- -zyme: to do with enzymes.
- -med or -medix or -medic or -medics: implies an application in the healthcare industry.
- -tech: superfluous, and obvious.
- -probe: either something to do with DNA probes, or something to do with medical diagnostics. Ideally both.
- -clone : suggests recombinant DNA technology.

The name can also have 'Sciences', 'Systems', or 'Technology' added to the end. If the name has several words in it, a memorable acronym can be helpful — hence DNAX, ABC, etc.

Neurotrophic factor

A general name for a nerve-specific growth factor, i.e. a molecule (usually a protein) which will encourage nerve cells to grow or to repair damage. Their potential use is as drugs to help patients overcome nerve damage caused by spinal or head injuries, degenerative diseases like multiple sclerosis or Alzheimer's disease, or ageing. Among the neurotrophic factors are:

- Nerve growth factor (NGF), the first neurotrophic factor to be discovered.

- Neurotropin-3 (NT-3), which is generating particular interest because it may have potential as a therapeutic for degenerative neural diseases such as multiple sclerosis or Alzheimer's disease.

- Ciliary neurotrophic factor (CNTF), which is believed to enhance nerve growth in embryos, and hence may be valuable for helping nerves to regenerate. Synergen are cloning this protein for eventual therapeutic use.

- Brain-derived neurotrophic factor (BDNF), which is similar to NGF but targets brain cells.

- Basic fibroblast growth factor (bFGF), which, in combination with NGF can help regeneration of central nervous system nerves in some animal studies.

New diseases

Because they have a panoply of new and powerful techniques at their disposal, biotechnologists are always looking for new ways of using them. One way is to identify a disease which has not been identified before, or one which is now thought to be much more serious than previously, and develop a treatment for it. For other diseases, of course, treatments already exist, which makes it harder for a new one to be accepted. Some of the 'hot' diseases discussed as targets for biotechnological solutions are:

- Any viral disease (as there are no effective anti viral drugs). Particularly AIDS (*see* **AIDS**), but also the following.

- Hepatitis, which is a degenerative liver disease (A,B,C viruses are well-characterized, with D and E being recognized, as well as environmental causes of the disease such as alcohol or solvent abuse).

- Herpes simplex, and particularly genital herpes, which can be dangerous to new babies if caught from their mothers. It is also an unpleasant disease for adults.

- Cytomegalovirus (CMV). A virus which causes glandular fever in children and adults, and is present in a 'latent' form in up to 60 per cent of normal people. It is not sufficiently severe to warrant a new treatment in most people, but can cause substantial disease in patients whose immune system is not working properly, especially AIDS patients.

Another disease in the news is:

- Lyme disease. A debilitating bacterial disease caused by the spirochete *Borrelia burgdorfei* first recognized in 1982 and now affecting thousands of people. A vaccine is wanted.

Nitrogen fixation

Nitrogen is an essential macronutrient (something we need a lot of in our diet) for all living things. Eighty per cent of air is nitrogen gas: however plants and animals cannot convert this into protein. They rely instead on other forms of nitrogen: ammonia and nitrates for plants, proteins and amino acids for animals. Only a few organisms can convert atmospheric nitrogen into these forms of nitrogen which can be assimilated readily, a process called 'fixing' the nitrogen. The rate at which fixed nitrogen can be supplied to growing crop plants is one of the limiting factors in their growth and yield.

Nitrogen fixing organisms are bacteria. Some live free in soil, some in symbiosis with plants. The symbiotic ones are the most interesting to the biotechnologist, although the free-living organisms like azobacter and klebsiella are easier to handle in the laboratory, and so most researchers prefer to use them. Symbiotic nitrogen fixing organisms live in nodules in the roots of a few plants, and convert atmospheric nitrogen to ammonia for the plants in return for a supply of C_4 acids, made by the plant from carbon dioxide. The genes which code for the enzymes which fix nitrogen — the *nif* genes — have been cloned and characterized in some detail. The *nod* genes, which induce the plant to make nodules in which the bacteria can live, are less well characterized, but the subject of intense study. Biotechnologists have tried several routes to fixing nitrogen for agriculture more efficiently.

Only a few types of crop plants (legumes, clover, rice, lupins) fix nitrogen via symbiotic bacteria of the genera rhizobium and bradyrhizobium, which live in their root nodules. Some other, non-legumes fix nitrogen, but are not widely used as crops. One route to making other plants capable of fixing nitrogen is to force the rhizobia to live in other plants, by introducing the bacteria to the plants in tissue culture or by engineering the cell-surface receptors of the plant root cells so that the bacteria absorb onto these roots in the same way as they do onto beans and clover. This route has been moderately successful on a laboratory scale. Another route, much touted ten years ago, was to **transfect** the *nif* genes into the plants themselves so that they did not need the bacteria at all. This is now thought to be an approach which is not likely to work. The bacteria provide a lot more enzymatic machinery than just the *nif* genes to fix nitrogen, and the roots also provide specific proteins (such as the plant haemoglobin protein, leghaemoglobin) which are essential parts of the nitrogen fixing process: the nodules are not just passive containers for the bacteria.

The simplest use of biotechnology is in producing legume inoculants, to increase the soil population of rhizobia around a growing legume. As each plant has to pick up the bacteria from the soil (there are no bacteria in the seed), nitrogen fixation can be limited by the rate of infection of the growing roots. Thus dosing the soil or coating the seeds prior to planting with suitable bacteria can give a better rate of fixation. (It is controversial as to whether this is an effective economic measure.)

A variant on this approach is to improve the efficiency of the bacteria in fixing nitrogen. BioTechnica tried an engineered *Rhizobium meliloti* in 1988, in which there were several copies of the gene for nitrogenase instead of the usual one copy. Nitrogenase is the enzyme which actually takes nitrogen molecules from the air and splits them open. The engineered bacterium was used to infect alfalfa, but as it did not result in any increased yield from the plant the trial was stopped.

If fixing nitrogen frees a plant from dependence on soil nitrates, why do all plants not fix their own nitrogen? The reason is that fixing nitrogen takes a great deal of metabolic energy, so if there is any other way to get nitrogen for the plant (or indeed for the bacteria), then they will take it as the more energy efficient course. It is not clear, therefore, whether getting plants that do not normally fix nitrogen to do so would actually decrease crop yields rather than increase them, as they would divert a lot of energy away from producing the edible portions of the plant and into nitrogen fixation, so leaving less for growth.

Oligonucleotides

Oligonucleotides are short DNA (or, rarely, RNA) molecules, usually defined as 100 bases long or less. This is the length of DNA which an automated DNA synthesis machine (a DNA synthesizer, oligonucleotide synthesizer, or 'gene machine') can make in one go and still have a significant yield of product. Oligonucleotides are usually defined by their origin — if it is made chemically it is an oligonucleotide, if it is cloned then it is a gene or a gene probe.

Oligonucleotides are usually named for their length. The naming follows the monomer – dimer – trimer scheme up to decamer (10 bases). Beyond that, the name of an oligonucleotide is generally its length *as a number* followed by '-mer'. Thus a 17-base oligonucleotide is called a '17-mer', pronounced 'seventeen mer'.

Automated DNA synthesizers use a series of chemical reactions to build up the DNA chain one base at a time. Each reaction consists of four steps, as the chemistry has to make sure that only one base is added each time, so building up a 50-base oligonucleotide (a '50-mer') requires 200 reaction steps. Clearly, if one of those steps is slightly inefficient, then the overall efficiency will be very poor — this is why synthesizing oligonucleotides of greater than 100 bases becomes quite difficult. Most gene machines are completely automated, so all the biotechnolgist has to do is type in the DNA sequence required and collect the DNA.

Oligonucleotides have become critical to biotechnologists for three reasons.

- They can be linked together to form larger lengths of DNA which can function as completely synthetic genes (*see* **Gene synthesis**).

- They can be used as DNA probes for a variety of genetic studies. In this they are particularly useful as they can distinguish between versions (alleles) of a gene which differ by only one base. Such oligonucleotides are called allele-specific oligonucleotides (ASOs).

- They are the primers for the widely used **PCR** technique.

Oncogenes

Oncogenes are genes which are believed to be necessary for cancers to develop. There are a large number of them and, as would be expected from the variety of cancer types, they act in many different ways. Most are present in normal cells as proto-oncogenes, i.e. versions of the gene which are benign, and indeed are essential to the body's normal development. A mutation turns them into the malign oncogene. There also exist anti-oncogenes (also called tumour suppressor genes), genes whose normal function is to suppress genetic activity which could promote the development of a cancer. If an anti-oncogene is mutated, then it 'releases' the activity of another gene and so hastens the development of the disease.

Oncogenes are of interest to biotechnologists because of the importance of cancer as a cause of morbidity and mortality in Western societies. Many biomedical research and development programmes are aimed at curing or palliating cancer, and hence are interested in directly or indirectly preventing the effect of oncogenes. The approach depends on the oncogene involved. Some oncogenes make proteins which are detectable on the outside of cells or in the blood: these proteins can be 'tumour markers', i.e. markers to show where a tumour is developing. In turn they may be used to diagnose cancer or to target a biotherapeutic to the cancer cell and so destroy it specifically. Oncogenes which act only inside the cells cannot be used as tumour markers in this way.

Among the more talked-about oncogenes are:

- *erb*, a family of proteins, of which *erb*-B2 is associated with breast cancer.

- *myc*: a protein in the cell nucleus, one of the earliest oncogenes identified (*see* **Oncomouse**).

- *fos*, a nuclear protein.

- *neu*, a membrane protein which is similar to the receptor for growth factors: it is believed that the mutant form 'looks' to the cell like a growth factor receptor which always has its growth factor bound, i.e. is always giving the cell a 'grow' signal.

- *ras*, a cell membrane protein which is associated with the protein kinase cascade, a complex series of enzymes which regulates many of the cell's functions in growth and differentiation.

- *tat*, a gene from HIV and many related retroviruses.

Many oncogenes have prefix letters. Thus there is c-*myc* (the cellular gene), v-*ras* (a viral, cancer-forming version of *ras*), H-*ras* (the human gene, to distinguish it from a number of homologues in other species.)

Oncomouse

An oncomouse is a semi-slang term for a transgenic mouse that has had a foreign oncogene placed into its genome. The first transgenic disease model, the oncomouse (or '*myc*-y-mouse') was developed at Harvard to model how one oncogene, the *myc* gene, contributed towards cancer. The gene is spliced together with the promoter from murine mammary tumour virus, which makes the gene express its protein specifically in the mammary gland. Rather than waiting for random mutation to turn the *myc* gene 'on', the transgenic oncomouse has a copy of the mutated gene already, and so develops mammary cancers at a very high rate. This in turn makes it a useful model both for detecting the other events which lead to cancer and for developing treatment strategies. Harvard subsequently patented the oncomouse, the first time an animal had been patented.

See also **Oncogene**.

Optical biosensors

A type of biosensor where the effect of a chemical on a biological system is detected using light rather than electrochemically. Several systems have been taken to commercial development in the last few years. They are based on the following principles.

Evanescent waves. When light is notionally trapped inside an optical fibre or a prism, in fact a little bit of it temporarily 'leaks out' to the outside world. The light outside its 'trap' is called the evanescent wave, because it is not 'really there' at all according to classical optics. If it finds a chemical there which can absorb it, then it is absorbed. Because the evanescent wave only occurs right next to the optical fibre or prism, it can only be absorbed by something stuck right on that fibre or prism. Thus measuring the absorbance of the evanescent wave allows us to detect when something has stuck to our optical surface *as opposed to* being free drifting around in solution.

If our optical fibre is coated with an antibody, then when that antibody captures its antigen that will change how the evanescent wave is absorbed, and so we can detect it. Variations on this idea have appeared as several near-commercial detection systems.

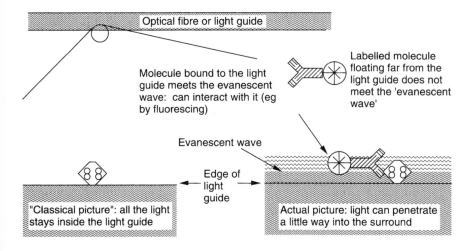

Optical fibre or light guide

Molecule bound to the light guide meets the evanescent wave: can interact with it (eg by fluorescing)

Labelled molecule floating far from the light guide does not meet the 'evanescent wave'

Evanescent wave

Edge of light guide

"Classical picture": all the light stays inside the light guide

Actual picture: light can penetrate a little way into the surround

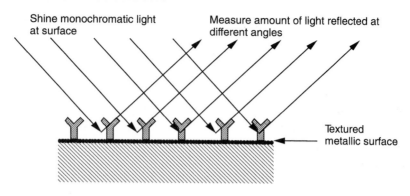

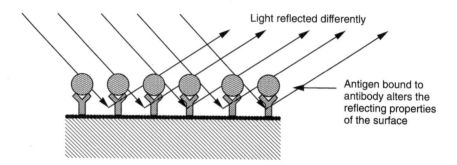

Surface plasmon resonance (SPR). This is a similar effect deriving from a different cause. When light is scattered off a conducting surface, the amount of light scattered at different angles depends on the exact nature of the surface and how it absorbs the light and conducts electricity. Thus if an antibody is stuck on the surface, how the surface reflects the light will vary depending on whether that antibody is or is not stuck to its antigen. Pharmacia have launched a commercial sensor system called BIAcore based on SPR.

The problem with all optical sensor systems has been that they give a lot of false signals, as anything which absorbs light can stick to them and give a 'positive' result. Thus the development work necessary to make them work reliably is not to get the optics to work *per se*, but to get them to work reliably in dirty biological samples. Many optical biosensor developments have foundered on this rock.

A lot of work has gone into making enzyme sensors work on optical fibres. Fibre optical chemical sensors (FOCSs) to measure pH, oxygen, and carbon dioxide are well-known, and have attracted a lot of interest for process monitoring and medical use because they are much stronger than ion-selective electrodes and, for medical applications, small enough to be inserted into a vein. The end of the optical fibre has a coating of a plastic which changes its optical properties when 'doped' with an ion, together with a chemical which selectively takes up just one ion into the plastic (an ionophore). Thus if that ion is present in the solution it is absorbed into the plastic, the optical properties (absorbence or fluorescence) change, and the detector 'looking' down the other end of the optical fibre can detect the change. Other ions are not absorbed and so are not taken up.

Biosensors seek to use this sensor approach by coupling enzymes to the end of a FOCS. When the enzyme causes a change in pH or consumes oxygen, the sensor detects this.

Organ culture

Organ culture is the growth *in vitro* of whole organs or parts of organs. Organs consist of several different cell types, as opposed to tissues which are made up of uniform cells.

In some ways organ culture is a part of 'traditional' transplant medicine. However some scientists are also developing artificial organ systems based on cultured cells on a synthetic matrix material which mimics the extracellular matrix of the body. Artificial skin is the most researched tissue: it can be made from dermis cells cultured in a suitable mesh of fibres, and has potential for use as a skin replacement for severe burns. Other potential tissue targets are vascular tissues, especially veins (as the active muscle in artery is harder to mimic).

A related area is bone marrow transplant, which is half-way between transplant and organ culture: here bone marrow cells are removed from one person and injected into another. However they are often treated to make them proliferate in between, and are sometimes subject to other treatments such as stimulation with specific cytokines or even to genetic manipulation.

Organic phase catalysis

This is the use of enzymes in liquids other than water. Organic phase catalysis (also solvent catalysis, hydrophobic catalysis, non-aqueous phase catalysis) is potentially useful for five reasons:

- The thermodynamics of the reaction may be more favourable in a non-aqueous solvent, giving better yields.
- The substrate may be more soluble in organic solvents (or indeed, only soluble in them).
- The enzyme may be more stable or have an altered specificity in the new solvent.
- There will be no side-reactions involving water.
- Products may be easier to recover from an organic solvent (e.g. by evaporation or extraction with water).

Thus, for some reactions, and especially those involving materials which are very poorly soluble in water or are very easily hydrolysed, getting an enzyme to work in a non-aqueous solvent could be a very good thing. Examples are the synthesis of peptides by proteases (in water, proteases only break peptides into amino acids) and the transesterification of lipids by lipases (in water, lipases overwhelmingly break lipids into fatty acids and glycerol rather than putting them together). Using lipases in organic solvents has been one of the more successful applications of this technology.

The problem is that, as usually prepared, enzymes rarely dissolve in anything except water, and even if they do dissolve they do not work. This is in part because enzymes are prepared as aqueous solutions, and so a mixture of the enzyme with an organic solvent is just that — a mixture of two immiscible liquids. If the enzyme is dried thoroughly so an absolutely minimal number of water molecules stick to it, some enzymes can be made to work in organic solvents such as octanol.

Variations include using supercritical fluids for the enzyme reaction, reversed phase or emulsion systems, or bioconversion in organic solvents.

An alternative approach is to genetically engineer the protein to be more stable or more active in the solvent concerned, and this is attracting some interest.

See also **Bioconversion in organic solvents, Lipases, Reversed phase biocatalysis, supercritical fluid enzymology**.

Orphan Drug Act

A US law which gives an incentive to a company developing a drug for a comparatively rare disease. For drugs which provide novel treatments for diseases which are suffered by only a small number of people, the Orphan Drug Act gives the developer of the first drug of any one type a 7-year exclusive right to market that drug. This is meant as an incentive to develop drugs for which the market would otherwise be marginal, given the intense competition in the pharmaceutical industry. It has been invoked quite widely by the biotechnology industry, as many biopharmaceuticals are so specific in their effects that they can only be used for a very narrow group of patients.

The Orphan Drug Act has come under attack recently for allowing biotechnology companies in particular to charge excessively for treatments for rare diseases. Because the act allows the companies a complete monopoly in the US, it is felt that some abuse their position. This is hotly disputed by the industry.

Osmotolerance in plants

Osmotolerance is a measure of a plant's ability to stand up to drought or to large amounts of salt in its water supply. Salt tolerance is sometimes called halotolerance. Because a reliable supply of fresh water is a limiting factor to agriculture in some places, osmotolerance is an important characteristic for plant breeders to achieve. Biotechnology has been suggested as a way of achieving it.

Plants survive water stress (i.e. environmental effects which tend to dehydrate them, such as drought or high salt) in a number of ways. These include structural adaptation (e.g. thickening cell walls to stop water loss, making leaves round to reduce the surface area), physiological adaptation (e.g. developing molecular pumping mechanisms to pump water into the cells or salt out), or metabolic adaptation (by producing internal chemicals which counteract the effect of drought or salt). Metabolic adaptations tend to involve only a few genes, where the other two involve many (dozens to hundreds). Thus the metabolic adaptations are the favourite target of biotechnological efforts to transfer osmotolerance to crop plants.

Metabolic approaches to osmotolerance usually involve filling the plant cell up with a compound which is innocuous, which the plant can make easily, and which 'attracts' water through its osmotic potential (i.e. just by being there, not because it expends any energy). A range of such compounds are known, and the enzymes to make them have been characterized to some degree or another. Consequently they could be genetically engineered into crop plants to make them capable of withstanding greater water stress. There are the usual problems of plant genetic engineering (e.g. Will it work? Will the resulting plant produce commercial levels of crop?), together with the additional problem that the osmoprotectant material must end up in the right part of the cell to be effective.

Oversight

In US regulatory contexts, this means 'having regulatory responsibility over'. Thus the definition of which organisms are subject to 'regulatory oversight' is a critical feature of the regulation of biotechnology — it defines which organisms must be approved by which authorities before they can be used for industrial biotechnology.

Patents

Whether a biotechnological process can be patented, and if so how, has been one of the biggest legal stumbling blocks to the technology's application since the early days of genetic engineering.

Nearly 23.5 per cent of all patents granted in Organization for Economic Co-operation and Development (OECD) countries in 1987 were granted in Japan. In the US 30.5 per cent in the FRG 8.8 per cent, and in the rest of the world less than 6 per cent for any one country. However Japan has a tradition of patenting everything (nearly 50 per cent of all *applications* are Japanese). Patents often form a type of trade barrier between countries, making it difficult for non-residents to obtain protection, and hence to use their invention in that country. For example, the US patent office has claimed that the Japanese patent system puts any foreign language applicant at a disadvantage.

The material which is patentable varies from country to country.

Area engineered	Macromolecules or viruses[†]	Unengineered microorganisms	Plant varieties	Animal varieties	Genetically-engineered organisms
USA	yes	yes	yes	yes	yes
Canada	yes	yes	no	no	yes
EPO*	yes	yes	no	no	yes
Japan	yes	yes	no	yes	yes

*EPO is the European Patent Office. The situation with the EPO is not clear. The ruling until recently was that it was not possible to obtain a European patent on a plant or animal. However it now seems that the EPO will accept patents on animals or plants, providing that they are formed as a result of 'a microbiological process'. The definition of a microbiological process is still unclear. † However, there is some uncertainty about what is the difference between a recombinant protein and its (presumably) identical natural counterpart, for example.

In addition to patents covering things ('composition of matter' patents), patents covering processes for making or using microbes are allowed in all the areas, although methods for breeding are not allowed by the EPO.

Apart from the differences and ambiguities in patent law, biotechnology companies face a substantially longer time between filling their patent and getting it granted than companies in most other fields, especially in the US. This means that they cannot defend their patent in the courts for several years after it has been made public.

Biotechnology companies have discovered that a patent is only as good as its last court case. While obtaining world-wide patent protection is complex and costly, the patentee has then to be willing and financially able to defend the patent against infringement in the courts, which can take years and cost millions of dollars.

Key organizations in patenting are: EPO (European Patent Office), PTO (US Patent and Trademark Office), and the various European national patent offices.

The most high profile patent cases in biotechnology are:

- PCR (polymerase chain reaction). There is no doubt that Cetus publicized and developed polymerase chain reaction. But did they invent it? Hoffmann LaRoche claims that they did not, and that it was described in 1973.

- Erythropoietin (EPO). Amgen and Genetics Institute have been working on genetically engineered erythropoietin roughly at the same time, and both tried to claim patent protection. In April 1991 the US Court of Appeal effectively gave complete rights to Amgen, because the supporting technical information provided with the Genetics Institute patent did not (the court said) enable someone else to reproduce what they did. ('Enablement' is a key part of a patent — the patent has to describe something new in a way which enables someone else to reproduce it.) This decision was a big surprise for industry observers, who had expected a 'mutual infringement' ruling, i.e. that they had both infringed each other's patents.

- Factor VIII. Factor VIII is used to treat haemophilia, and Genentech, Scripps Clinic, and Chiron had developed methods for purifying it from blood, and had claimed a patent on the product. The US Appeals Court decided that they could not claim rights to the product (although their specific ways of making it are patentable).

- cDNA. Recently Craig Venter at the NIH in the US has filed a patent claiming the sequence of 337 cDNA clones, copies of naturally occurring RNAs. If this filing is passed by the patent examiners in the US, then the NIH will be able to claim royalties from anyone who subsequently finds out what the cDNAs code for and whether any of them are of any use. Supporters of this approach say that other people have patented their cDNAs before, so why should Venter not patent his simply because he is more efficient at sequencing them? Opponents say that he has invented nothing new — he does not know what proteins the cDNAs code for, nor what could be done with the cDNAs or with the proteins they code. The US patent examiners' preliminary decision rejected the application, a decision that is being appealed.

See also **Blood disorders, cDNA, Growth factors, PCR**.

PCR (polymerase chain reaction)

Polymerase chain reaction is a method for amplifying DNA that is generally believed to have been invented by Kary Mullis of Cetus (*see* **Patents**). It takes a single copy of a DNA molecule and uses it to create millions or billions of copies of itself. Because of the specificity and accuracy of the reaction, this is an ultimately sensitive detection system, able to detect one molecule in a reaction.

The diagram outlines how PCR works. The key ingredients are Taq polymerase (DNA polymerase, an enzyme which makes new DNA) isolated from the bacterium *Thermus acquaticus* or some other, equivalent heat-stable DNA polymerase, and two primers, short DNA molecules, which are complementary to two sites either side of the piece of DNA you want to amplify. The primers are usually oligonucleotides which one has synthesized. Given those two ingredients, PCR will amplify almost any piece of DNA.

Many uses for PCR have been developed since its invention in 1985. Most obviously it has been used for detecting DNA sequences, for the diagnosis of genetic disease, for DNA fingerprinting (*see* **DNA finger-printing**), for detecting bacteria or viruses, and for research (especially such arcane matters as cloning the DNA from Egyptian mummies and from dodos). Its use in genetic diagnosis is widespread, while the use in bacteriology is much less common. This is in part because of the problem of contamination. If PCR can amplify a single molecule of DNA, then a single molecule 'escaping' from the amplified product, if it gets back into the starting materials, can start the PCR reaction. Several workers have had to abandon research into a particular gene because their laboratories have become saturated with contaminating PCR products, and some genetic diagnoses detecting male-specific genes in fetuses have to be done exclusively by female workers as the skin cells falling of male workers are enough to contaminate the test.

PCR can also be used to clone genes, if two suitable primers can be made, and to select the correct gene construct from a mixture of constructs when making a synthetic gene: the use of PCR in cloning is very widespread.

Variants on PCR such as single-sided PCR (which rearranges the DNA before amplification so that only one primer is needed), inverse PCR (which also rearranges the DNA, this time to amplify the DNA that flanks two primers, rather than that which lies between them), and random PCR (which stitches synthetic DNA onto the ends of a segment to be amplified so that no new primers are needed) have been developed.

PCR is the subject of a bitter patent dispute between Cetus, who claim they invented it, and Hoffmann LaRoche, who say that it was essentially

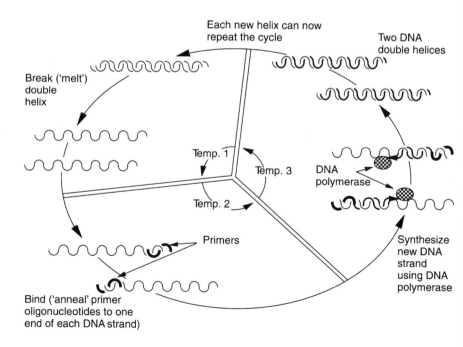

invented fifteen years before. Partly because of this and partly because the Cetus patent covers all applications of PCR, there are a number of other amplification systems which do similar things but which work via a different mechanism.

See also **DNA amplification**.

Peptide synthesis

Peptides are very short strings of amino acids, usually ten to twenty amino acids long, but sometimes only two or three. They are made by different routes from proteins, for two reasons. Firstly, peptides are usually broken down rapidly by bacterial cells, so it is difficult to make them by recombinant DNA methods. Secondly, because they are relatively small molecules, it is feasible to make them by chemical or enyzmatic methods.

There are three general routes for making peptides. The first is by genetic engineering. The peptide is usually produced as a fusion protein, the peptide itself being joined onto a much larger protein. It then has to be cut ('cleaved') from the larger entity after that has been purified from the bacterium or yeast that has made it. This can be difficult to achieve effectively, as you need a chemical reagent (such as cyanogen bromide, which cuts at methionine residues) or an enzyme which cuts the fusion protein at exactly the junction between the peptide and the larger protein, but not within the peptide itself.

The second route is *in vitro* enzymology. Many proteases which break peptide bonds are known. By altering their reaction conditions they can be made to work in reverse and synthesize peptide bonds. Conditions can include making them work in organic solvents (*see* **Organic phase catalysis**), under extremely high pressure, or modifying the amino acids so that the peptide is removed from the reaction (by precipitation or because it dissolves in a second, organic solvent phase) as soon as it is formed.

To prevent the protease joining together a whole string of amino acids, but rather add them one at a time, the amino acids are 'protected' by adding groups to them which prevent uncontrolled polymerization. A cycle of reactions adds an amino acid, then removes its protecting group, then adds another and so on.

The third route is chemical synthesis. This performs the same sort of reaction cycle as enzymatic synthesis, but uses traditional organic chemical reactions. The reactions can be carried out on a solid material (in a reaction series called the Merrifield synthesis) so that the peptide chain 'grows' while attached to a support structure, or in solution, which is usually easier for large amounts but cannot make long peptides. The efficiency of each step is high, but because it is not 100 per cent the yield is usually low after a couple of dozen amino acids have been added.

Chemical routes usually require more reaction steps than enzymatic ones, but the materials are usually cheaper. Either enzymatic or chemical synthesis can produce kilograms of a peptide, and there are fully automated 'peptide synthesizers' which can perform the chemistry to synthesize grams of a peptide in a few hours.

Peptides

Peptides are short protein molecules, but usually are produced using methods different from those used to make other, longer proteins. In general, something is a peptide if it contains 20 amino acids or less, a protein if it contains 50 amino acids or more: in between, it depends on who you ask.

Peptides were very popular in the 1980s, as it was discovered that a large number of hormones and neurotransmitters (the hormones which carry signals between nerve cells) are peptides. They can be produced by chemical, biochemical, or genetic means, unlike larger proteins which are usually produced solely by genetic or cell biological methods. Chemical synthesis adds amino acids one at a time to a growing chain using a cycle of reactions.

Peptides which have been made commercially include calcitonin (for osteoporosis), glucagon (for hypoglycaemia), thyrotropin-releasing hormone (TRH, for thyroid disease). Aspartame, the artificial sweetener marketed under the Nutrasweet label, is a two-amino acid peptide, and is produced in amounts that dwarf the other, pharmaceutical products (*see* **Artificial sweeteners**).

See also **Peptide synthesis**.

Permeabilization of cells

Cells are normally surrounded by a thin membrane of lipids and proteins — the plasma membrane. This is meant to keep out anything not essential for the cell's survival (and, for cells from animals and plants, their function as part of the whole). However they can also keep out materials that biotechnologists want to get in. To get round this, cells can be 'permeabilized'. This effectively makes small holes in the plasma membrane, so that material can get into the cells but all their contents do not leak out: they remain capable of doing what is required.

Permeabilization can be done by treating the cells with organic solvents (which dissolve out small patches of the lipid membranes), detergents such as bile salts, some special-purpose ionophores (molecules which introduce channels of molecular size into the membrane, and which usually only let in a limited number of types of molecule) or physical treatments like **freeze-drying** or sonication (exposing the cells to intense ultrasound). Many types of cell also become rather more permeable to some chemicals after they are immobilized onto solid supports.

Permeabilized cells can have other advantages over intact cells for use in a **bioreactor**. They are rarely viable, so they do not waste metabolic energy (and hence your valuable starting materials) building more cell mass. They will also not grow inside a bioreactor, blocking it up.

Pest resistance in plants

As a potential alternative to using conventional pesticides, genetic engineers have sought to introduce genes conferring resistance to pests into plants. There are two potential routes for this.

The first is to identify existing genes in plants which confer resistance to pests, and transfer them to crop plants which are more valuable but are susceptible to that pest. This route is favoured in searches for resistance to bacterial and fungal pathogens. The plant genes often show 'gene-for-gene' matching with genes in the pathogen called 'avirulence genes': the avirulence genes have a role in causing the disease, and the corresponding plant genes have evolved to stop them. The difficulty is that exactly *what* the genes are doing is often unknown.

The other route is to add a completely new gene to the plant. This is a way to fight off pests which will not respond to changes in plant biochemistry, usually pests which do gross damage to plants by eating them. The approaches currently used are:

- To include the gene for *Bacillus thuringiensis* toxin in the plant. The toxin stops gut function in some insects, so that if they nibble the leaf it kills them. Calgene have successfully done this with tobacco, and Monsanto with tomato — the latter was a dramatic success as far as the plant's resistance to insect pests was concerned. Plant Genetic Systems have a number of field trials of plants engineered with *B.t.k.* toxin in Europe and the US, including potatoes and tomatoes, and Sandoz Pharmaceuticals is commercializing their transgenic *B.t.k.* toxin tobacco in the US. As tobacco is grown to be burnt, not eaten, there are fewer concerns about the health safety of genetically engineered tobacco than about almost any other crop.

- To include an enzyme which attacks insects in the plant. DNA Plant Technologies is working on this, using chitinase as the enzyme: chitin is a major component of insects' skeletons, and chitinase is an enzyme that breaks it down.

- To include a protein which blocks a pest's usual method of attack or digestion of the plant. This has been used with good effect: the gene for cowpea trypsin inhibitor, a protein which inhibits the protease trypsin (and related enzymes), has been engineered into tobacco. This blocked the action of digestive enzymes in insects' guts, and so killed them. Chitinase also works through this route to an extent, breaking down the wall of the gut.

See also **Biopesticide**.

Pharmaceutical proteins

Pharmaceutical proteins, also often called biopharmaceuticals and sometimes also 'biologics' (in regulatory contexts), are proteins made for use as drugs. Some of the most highly publicized applications of biotechnology have been in the production of biopharmaceuticals, and indeed the earliest generally recognized products of the current wave of biotechnology products — somatostatin and human insulin — are biopharmaceuticals.

Usually biopharmaceuticals, which have to be human proteins to be fully effective in humans, are made from genetically engineered bacteria, as the only other source is cadavers or live human tissue. The genetic engineering of such products is covered elsewhere. Issues peculiar to biopharmaceuticals are usually the result of the stringent regulation that any drug must pass before it is allowed to go into general use. They include:

- Demonstration of efficacy. A curiosity of the regulations is that each biopharmaceutical must be demonstrated to be effective on its own, whereas many of them are meant to be adjuncts to therapy with other drugs and not to have an effect of their own.

- Demonstration that the product is free of contaminants. This is particularly true of bacterial proteins and cell wall material which could act as a 'pyrogen', i.e. a material which could cause a feverish immune response in someone injected with it.

- Demonstration of purity and stability. There can be materials other than the biopharmaceutical in the preparation — indeed, some are so powerful that a single does of a few milligrams would not be visible to the naked eye, so something else has to go in there just to make it easier to handle. However the 'something else' must be exactly characterized. The whole preparation must also be demonstrated to be stable. This is often achieved by freeze-drying it.

- Freedom from side-effects. Apart from those caused by impurities or extremely high doses, these include principally the body's ability to recognize the protein as foreign, and so mount an immune response against it. Differences as small as the removal of an N-terminal methionine from a protein can alter the bodies immune response to it.

See also **Drug development pathway**.

Pharmacokinetics

This is the study of how a drug's effective concentration changes over time. The amount of a drug in the body depends on how much was given, how fast it is broken down, and how fast it is excreted. The speed of breakdown is a particularly crucial point for biopharmaceuticals, as many recombinant proteins are liable to be mopped up by the body's immune system or by the normal mechanisms which remove 'old' proteins from the blood. Altering the glycosylation patterns of recombinant proteins can later their pharmacokinetics substantially, which is one reason why glycosylation patterns are important for biotechnological drug producers.

Physical containment

Physical containment of genetically engineered organisms is the principle way in which they are kept inside a laboratory and prevented from 'escaping' to the wider world. (The other route is biological containment.) This is containment by physical barriers. There is a range of physical barriers used, many of which are similar to those used in building clean rooms: however in a containment laboratory the idea is to keep the dirt in, not out.

Air filtration. Exhaust air from air conditioning is filtered before it is vented to the outside. Often the containment laboratory is kept at a lower pressure than the outside ('negative' pressure) so that any air leaks leak into the laboratory, not out.

Sterilization lights, usually versions of fluorescent tube lighting which give out a lot of ultraviolet light, is commonly used to sterilize the laboratory's exposed surfaces at night (when it will not give workers sunburn).

Waste disposal. Often all waste leaving a containment laboratory is autoclaved to sterilize it. This includes apparently innocuous waste like paper towels as well as obviously contaminated material. An alternative is to incinerate it, but then it has to be sealed to take it to the incinerator.

Personnel protection. Personnel working in a containment laboratory must often wear protective clothing, much as they are required to do in a clean room. This, however, is so that potentially contaminated clothing can be thrown away on leaving the room and not carried into the outside world.

National governments define several levels of containment under which different procedures have to be carried out. Typical levels would be:

- Level 0: any laboratory.

- Level 1: 'Good microbiological practice'. This is equivalent to any microbiology laboratory, where normal microbiological techniques are used to ensure that relatively non-hazardous organisms are kept in the lab, and do not cross-contaminate experiments. Typically such laboratories are used for routine gene cloning involving no expression of a gene which could be hazardous to humans.

- Level 2. The laboratory is kept at negative pressure and air filtered. Any contaminated waste is autoclaved. Initial gene cloning experiments

PHYSICAL CONTAINMENT

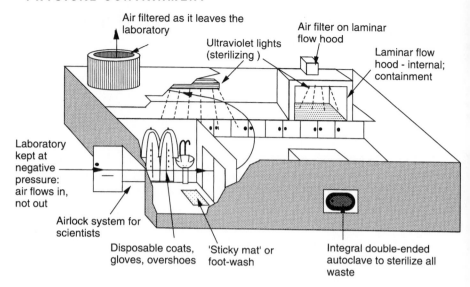

Air filtered as it leaves the laboratory

Ultraviolet lights (sterilizing)

Air filter on laminar flow hood

Laminar flow hood - internal; containment

Laboratory kept at negative pressure: air flows in, not out

Airlock system for scientists

Disposable coats, gloves, overshoes

'Sticky mat' or foot-wash

Integral double-ended autoclave to sterilize all waste

involving high levels of protein expression may be carried out in such laboratories, as well as microbiology involving organisms which have a relatively low hazard risk. As an additional safety precaution, most work would be done inside laminar flow hoods, hoods in which the air is circulated so that any particles generated by the experiment are carried up into the hood's own filter system and not into the laboratory.

- Level 3. The laboratory is only entered through an 'airlock' system, and all waste leaving it is autoclaved. Workers have to wear elementary protective clothing. Work on genetically engineered organisms which are expressing bioactive proteins, and on dangerous but relatively non-infectious organisms such as clostridia would be done in such laboratories.

- Level 4. This is the ultimate containment level in most countries. Air is usually double-filtered on the way out, there is a double-airlock system for personnel with disinfectant bath to wash their shoes/boots in on the way out, and no one is allowed into the laboratory without substantial training (and no one who does not need to be there). Work on 'live' AIDS viruses and genetic engineering of common bacteria to express highly toxic proteins such as ricin could be done in such facilities.

Level 4 facilities are rare: usually even the most potentially hazardous biotechnology project is adequately contained by Level 3.

See also **Biological containment, Clean room, GLP/GMP, Sterilization**.

Plant cell culture

Like any living organism, plants are composed of cells, which are capable of growing and dividing outside the plant given the right conditions. However, these conditions are rather specialized, as the plant cells are themselves specialized to work most effectively inside a plant. Thus the conditions for cell culture have to provide the cells with a range of nutrients and, most essentially, keep the cells free from any other contaminating organism like a bacterium or fungus. Although plant cells have a range of defences against infection, the bacteria or fungi can grow very much faster than the plant cells in fermenters, and so outgrow the plant cells, resulting in a large mass of contaminant and either a small mass of plant cells or the death of the plant cells.

Plant cell culture has a wide range of applications in biotechnology, in

- Plant cloning, i.e. the growth of plants from very small pieces of plant tissue, even single plant cells (*see* **Plant cloning**).

- Plant genetic engineering (*see* **Plant genetic engineering**).

- Making 'plant' products (like scents or food flavours) from plant cells in culture rather than whole plants. Plants produce a very large number of useful chemicals, but often do so only at certain times of year and in places where growing the plant is difficult or even dangerous. Ideally, if the cells from the plant could be grown in a **bioreactor**, then some of these inconveniences could be avoided. The problems arise primarily from the way that plant cells in culture produce very little of these secondary metabolites. This can be overcome in some cases by growing the cells with suitable elicitors, compounds or mixes of compounds (often from plant of fungal sources) which are observed to increase the rate of production of secondary metabolites in cultured cells.

In this the plant biotechnologist is helped by the plant cell's totipotency. Most plant cells are capable of being grown back into a whole plant — they are totipotent, i.e. they have all the 'potency' of the original plant. This contrasts to animal cells, most of which cannot be grown into anything other than the tissue from which they came.

See also **Embryogenesis, Secondary metabolites**.

Plant cell immobilization

As well as the general methods used to immobilize growing cells in a **bioreactor** there are several techniques which are relatively specific for immobilizing plant cells.

Entrapment of plant cells in gel matrices in popular: the cells are suspended in small drops of the material, which then set or harden to make little carriers. Materials such as alginates, agar, or carageenans (all of which are polysaccharides from seaweed), gelatin, or polyacrylamide have been used. **Hollow fibres** have been used for plant cells, but are not as popular as for animal cells, in part because hollow fibres are ideal for maintaining cells which are secreting some produce, and few plant cells secrete anything in worthwhile amounts. A relatively new method involves immobilizing the cells in polyurethane foam. In these foam reactors small lumps of foam are suspended in culture media and the cells encouraged to grow into the holes inside, where they form mini-bioreactors.

Unlike animal cells, plant cells are enclosed in a very tough cell wall. This means that plant cells will not spontaneously stick to a substrate as easily as animal cells will. However you can chemically link them to one without killing them. Plant cells have been chemically linked to nylon and to polyphenylene beads using glutaraldehyde (a standard chemical for linking two biopolymers together).

See also **Animal cell immobilization**.

Plant cloning

One area in which traditional biotechnology has been successful is in plant cloning, based on the techniques of plant cell culture and embryogenesis. The technique is an extension of the idea of taking cuttings of a plant to 'duplicate' a particularly valuable plant. With cell culture techniques, the 'cutting' is a single cell.

Cloning from plant cells involves several steps.

- Isolating individual cells. If all you want is a number of plants, then the cells need not be rigorously separated from each other: if not, they can be small chunks of tissue (tissue explants).

- Genetic manipulation of the cells.

- Callus generation: culturing the plant cell into a mass of cells, looking like a small piece of chewed paper.

- Embryogenesis/organogenesis: the callus is encouraged to regenerate roots and leaves.

- Planting: once the plant cells have generated a recognizable plant, it is safe to put it in the soil and see if it will grow.

A further step is the use of anther cultures to speed up breeding programmes for obtaining homozygous plant lines (that is, plants in which both copies of all genes are the same, so they breed true for all traits). Anthers from male plants are cultured, and the haploid cells (i.e. the cells containing only one set of chromosomes, not the normal two) in the anther encouraged to grow clonally into plants. Unlike animals, haploid plants are often capable of growing in culture. As they only have one set of chromosomes, on 'diploidization' (i.e. any technique which will double up their chromosomes to make a normal, diploid plant) both copies of their chromosomes will be the same, i.e. they will be homozygous.

There are two major problems with using this sort of technology routinely for propagating plants. Firstly, the conditions for getting the callus to grow and then to differentiate are different for each plant. It is largely a matter of trial and error whether the right combination is found for the species in question. Secondly, plants have very effective ways of fighting off parasites like fungi and bacteria. However these defences are much less effective in culture. Thus the initial plant material must be sterile before you start the culture, which is very hard to achieve for something that spends 24 hours a day sitting in the soil.

A third problem of somaclonal variation arises in some species. If a potato is separated into its component cells and some of these regenerated into potato plants, few of them will turn out to be identical to the original plant. This is a genetic change, a reflection of genetic instability. This is not a feature of the whole plant, which may be bred using normal methods quite well, and so must be an effect of the cell culture system.

Why it happens is not understood, but it is one reason why some plants are not cloned in this way.

See also **Embryogenesis, Plant cell culture, Plant genetic engineering, Somaclonal variation**.

Plant genetic engineering

Plant genetic engineering is a major part of research effort in biotechnology, because of the potential it holds out for improving crop plants. A genetically engineered plant, sometimes called a transgenic plant, is the product of several technologies covered in this book. The necessary steps to make a transgenic plant are:

- isolating single plant cells (*see* **Plant cell culture**);
- getting DNA into those cells;
- regenerating the cells into plants again; and
- in some cases making homozygous plants from heterozygous transgenics (*see* **Embryogenesis, Plant cloning**).

Getting DNA into plant cells has been difficult, because plant cells are surrounded by a robust cell wall and, unlike bacterial cells, do not have common mechanisms for acquiring DNA from their surroundings. As with all methods of making truly genetically engineered multicellular organisms, the key is not only to get DNA into the plant but to get it in a suitable amount and to have it integrated into the plant's chromosomes.

The common routes discussed are:

- Using *Agrobacterium tumefaciens* (*see* **Agrobacterium tumefaciens**).
- Using electroporation on plant protoplasts (*see* **Electroporation**).
- By microinjection. This technique, which has worked so well in creating transgenic animals, has been applied to plants in two ways. Plant cells have been injected with liposomes containing DNA. Providing the

liposomes are not injected into a vacuole, this is an effective way of transferring DNA into the cell. The alternative microinjection route is to inject DNA directly into the nucleus of the cell. This is more difficult to do, but gives greater control over the amount of DNA injected.

- By biolistic (particle gun) delivery. This is a favoured route, and is efficient at getting DNA into plant cells. However the DNA is only integrated into the plant's chromosomes with a low efficiency, so this is a relatively inefficient way of making transgenic plants (as opposed to just getting DNA into plant cells for research study, *see* **Biolistics**).

- By transformation of protoplasts. If the plant cell's wall is removed, then the resulting protoplast can sometimes be transformed simply by mixing them with DNA (in the right conditions). This has not worked with monocotyledons yet (most major crop plants like wheat and maize are monocotyledons), and seems to have only limited potential (*see* **Protoplasts**).

After a gene has been introduced into a cell, the one cell among many thousands or millions which has taken up the gene must be identified. This is the selection stage of genetic engineering, and as with bacterial or yeast genetic engineering usually relies on a selectable gene which you have transfected into the plant cell together with the gene you want in there. This gene may be for resistance to a herbicide (which would kill the plant cell), or for an enzyme which is easy to detect using a simple assay (so you can look through your plant cells for ones which have that enzymic activity). You can also screen cells for the presence of DNA itself using hybridization. This is more difficult to do with plant cells than with most other types of cell, because plant cells contain relatively little DNA (compared to yeast or bacterial cells) and it is quite hard to release.

Potential targets for plant genetic engineering fall into a limited range of types of project.

- Pest resistance, engineering genes into plants which will help them to repel pathogens.

- Herbicide resistance, putting the genes for herbicide resistance into crop plants so that they can be resistant to the herbicides which kill weeds.

- Nitrogen fixation, using a variety of routes to make plants 'fix' nitrogen from the air instead of needing fertilizers.

See also **Nitrogen fixation, Pest resistance in plants**.

Plant oils

A substantial part of commercial biotechnology is aimed at producing or modifying plant oils. The oils are stored in the plants as triacylglycerols (TAGs), i.e. molecules with one fatty acid linked to each of the three hydroxyls of glycerol.

Common sources of oils include palm and coconut (medium chain oils), used mostly in detergents; rapeseed (long chain oils), used as lubricants, plasticizers, and for making nylon; castor bean and lesquerella oil (hydroxylipids), used in lubricants and coatings; jojoba wax used in lubricants and cosmetics; flax oil (trienoic), used in coating, drying agents, and to a small extent in cosmetics; and cocoa used in chocolate and cosmetics.

Enzymatic processes involving the use of plant oils include hydrolysis (to make the fatty acid) and transesterification (to make different esters from the glycerol and fatty acids).

See also **Lipases**.

Plant sterility

An important aspect of plant breeding programmes is obtaining genes which confer sterility. This is in part so that farmers cannot breed from the seeds they are provided with, in part to assist breeding programmes, but mainly so that hybridization breeding methods can work. These produce 'hybrid' grain crops, i.e. crops in which the seed you plant is the result of crossing two other types of grain plant. The two parental strains do not themselves produce high-quality grain. They produce grain which grows into a high-quality crop. This enables characteristics to be combined into one crop plant which could not be maintained by the traditional practice of keeping back a fraction of this year's crop to plant next year.

However it is essential that the grain sold to the farmer is the offspring of the mating of both parental types, and not just one. This requires the breeder to select male plants from one type and female plants from another — as sexing a field of wheat is tedious, this is done by ensuring that the various combinations you do not want are sterile, i.e. set no seed. Usually it is the male plant which is sterilized, and so the genetic effect is often called 'male sterility'.

Biotechnologists have provided a range of new ways of making plants sterile, either one sex or both sexes. They have also generated 'restorer genes', genes which reverse the effect of the male sterility gene. This allows the plants carrying the male sterility gene to be cultivated on their own — without it, the line of plants would die out within one generation because of the lack of males.

Plant storage proteins

Plant storage proteins are proteins accumulated in large amounts in seeds, not because of their enzymic or structural properties but simply as a convenient source of amino acids for use when the seed germinates. They are of interest to biotechnologists for two reasons.

Storage proteins as a source of protein. Much of the world's food comes from plant seeds or fruits, and much of the protein in those seeds is storage protein. Thus a substantial amount of the world's food protein comes from plant storage protein. Any improvement of the nutritional content of those proteins could correspondingly improve human diet. Specifically, many storage proteins are poor in some essential amino acids, usually the sulphur-containing ones. They are called class II proteins, because they cannot provide a good source of protein for humans on their own. A diet which relies on just one storage protein source for nearly all its protein can be deficient in one or two amino acids, despite being quite adequate in bulk protein, and lead to deficiency disease. Improvement of the proteins for food use would seek to engineer them to contain more of the essential amino acids, and so be better balanced, class 1 sources of protein.

PLANT STORAGE PROTEINS

Storage proteins as expression systems. Storage proteins are produced in very large amounts relative to other proteins, and are stored in stable, compact bodies in the plant seed. Several workers are seeking to make the plants produce other proteins in similarly large amounts (up to 60 per cent of the total seed protein, 15 per cent of the total seed weight) and in as convenient a form. Storage proteins are often **glycosylated** as well, although often not in the same way as a protein would be glycosylated by mammalian cells.

The favoured route, being tried by Plant Genetic Systems, is to splice the gene for a desired protein into the middle of a plant storage protein gene. This construct will then produce a fusion protein in the seeds, which can be chopped up to yield the desired product afterwards. The favoured protein to do this is the small 2S plant storage protein, and has been achieved with a model system in *Arabidopsis thaliana* and in *Brassica napus* (oilseed rape). This may not be the ideal protein, as, because it is small, splicing a large gene into the middle of it could disrupt its structure.

A more radical approach would be to use the promoters of a storage protein to make a completely synthetic gene. This could be very difficult, as, if the protein is not simply to be destroyed, it must also be targeted to the storage vacuoles in the seed. The targeting mechanism for seed storage vacuoles is not known, although proteins have been targeted to the vacuoles of other plant cells successfully.

Plasmid

A plasmid is a small piece of DNA which can exist inside a cell separate from the cell's main DNA. This means that it must be able to replicate itself inside the cell, so plasmids have the correct genetic elements in them to cause the cell's enzymes to replicate them as the cell divides.

Plasmids exist in most microorganisms. Those in bacteria are almost invariably circles of DNA. Some in yeast are linear DNAs, like very small chromosomes.

Plasmids are used extensively in genetic engineering as the basis for vector molecules. Because they are small, they are easy to manipulate. (By contrast the chromosome of *E. coli*, with three million bases, is a molecule 2 nm thick linked into a circle 1 mm in circumference. A tube with a billion of these in is too thick to pour, and the shear forces of stirring it will break most of the molecules). Plasmids also have few sites for restriction enzymes in them, and so it is relatively easy to cut them open at just one place, then to splice in a piece of 'foreign' DNA and join the ends up again. They can also be manipulated to be present in many copies in the cell, rather than the one copy of normal chromosomes and plasmids.

Plasmids are a specific type of episome, the generic name for any small DNA that can exist as an independent entity inside a cell free of the cell's main chromosomes. Some viruses can also be episomes, existing as DNA within a cell for a long time. (This does not include the **retroviruses**. These exist as DNA inside a cell, but their DNA is spliced into the chromosomes themselves.)

See also **Vector.**

Polysaccharide processing

One of the most common uses of industrial enzymes is in the food industry, and particularly in the processing of complex polysaccharides such as starch and pectins. Enzymes are involved in several processes.

Liquefaction. The dispersal of starch granules into a gelatinous suspension (i.e. essentially what happens when cornflour is boiled to thicken sauces). The starch is also hydrolysed into shorter molecules by enzymes such as pullulanase and alpha-amylase. Because liquefaction is often carried out in hot solutions, one valuable biotechnological product is heat-stable alpha-amylase and pullulanase, isolated from thermophilic bacteria, which will work at 80° or 90°C.

Saccharification. The formation of low molecular weight sugars, often mainly glucose, from liquefied starch. There are a variety of enzymes which will do this: amylases and pullulanases to break down the starch, invertase to break down sucrose, and glucose isomerase to convert glucose into the sweeter fructose.

Debranching. A chemical rather than a process term, this is the removal of the side branches from the long starch or pectin molecules, leaving long, straight molecules which are easier to break down further. Branched and unbranched polysaccharides also have different gelling properties, and so confer different mechanical properties on food. Enzymes such as pullulanase and isoamylase can perform debranching of starch

See also **Glycosidases.**

Post-translational modification

This is a blanket term to cover the alterations that a protein undergoes after it has been synthesized as a primary polypeptide. They include the following.

Glycosylation. This is one of the critical post-translational modifications for biopharmaceuticals (*see* **Glycosylation**).

Removal of the N-terminal methionine (or N-formyl methionine). Nearly all proteins are made with a methionine as their first amino acid, and this is usually removed. Sometimes it is removed as part of:

Signal peptide removal. Peptides which are to be inserted into membranes, secreted into special cellular compartments (like the mitochondrion or into vacuoles or lysosomes) have a short string of amino acids at their front called the signal peptide. This signals to the cell where the protein is to go, and is chopped off as part of the mechanism for getting it there.

Acetylation, formlyation. These and a few other modifications put relatively unreactive groups onto more reactive ones. They are often put on the terminal amino group of a protein, producing a 'protected N-terminus'.

Amino acid modification. This is the chemical modification of amino

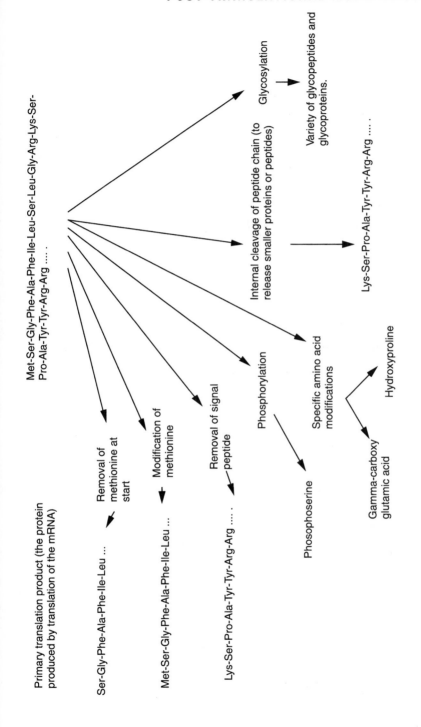

acids after they have been incorporated into the protein chain. It is relatively rare, but can have critical effects on the protein's function. Examples are the modification of glutamate to form gamma-carboxy glutamate by a vitamin K-catalysed reaction in mammalian liver, and the hydroxylation of proline to hydroxyproline in collagen in animals.

See also **Expression systems, Secretion.**

Predisposition analysis

This is the analysis of how some people are more likely to get some diseases as a result of their genes. Many diseases have a 'genetic component' and an 'environmental component', and a 'bad' environment or a 'bad' gene can enhance the chances of contracting the disease. For some rare diseases of the immune system such as ankylosing spondylitis there is an 80-fold higher chance that the carriers of some genes will get the disease than the carriers of others. For other diseases the effects are less dramatic. Among those discussed as having a genetic component are:

- many immune disorders, including asthma, eczema, Graves disease, allergies;
- diabetes;
- hypertension;
- some types of cancer (but not most cancers);
- hypersensitivities and adverse reactions to drugs and chemicals.

And a range of other diseases may have a substantial genetic component, for example:

- schizophrenia;
- clinical depression;
- cardiovascular disease.

The biotechnological interest in this genetic predisposition is threefold.

Firstly, if there is a gene involved, we could hope to use DNA technology to detect that gene and so detect who is predisposed to that disease. Secondly, we could hope to find out what the gene does and hence design a therapy to counteract it. Lastly, we could also try to identify the environment which acts with the gene to cause the disease, and hence reduce the disease's incidence by reducing the chance that *anyone* will be exposed to that environment.

There are obvious ethical and legal implications in using human genetic information in this way. However there are also practical implications. Most of these predispositions will not be caused by 'a gene', but a number of genes, all of which will have to be characterized and understood. In addition, the effects of the genes will not be obvious in everybody — they will predispose to a disease, not inevitably cause it. This means that they can only be identified from large statistical surveys. This is a major research undertaking, and this is one of the reasons why, when the genes for many rare genetic diseases have been found, the gene(s) for the much more common disease such as hypertension are still unknown.

Notwithstanding this, several companies have been set up to use DNA techniques to detect predisposition to disease, and one of the aims of the Human Genome Project (*see* **Genome project (HUGO)**) is to provide information about genes which could predispose people to some diseases.

Proteases

Proteases are enzymes which break down proteins. There are four distinct uses for proteases in biotechnology. Their use depends partly on how cheap they are to make, and partly on how specific they are — i.e. whether they chop up all proteins indiscriminately or only a few proteins at specific points.

Eight thousand tonnes of proteases from microbial and fungal sources are made each year, and most of them are used in detergents. Relatively non-specific proteases are used to digest the protein material in dirt — it is often the denatured protein which makes organic stains hard to wash out. Some such detergents are retail products, but more are used in industrial cleaning, as the proteases are powerful enzymes and can strip the protein out of the user's skin if not handled carefully.

Their other main use is in the food industry, where microbial rennin is used extensively in cheesemaking as an alternative to cows stomach rennin. A rising trend is the use of proteases to tenderize meat and to enhance the flavour of foods by altering the proteins in them. This use requires purer proteins (as they, or their cooked remains, are going to be eaten), and they are usually quite specific, only cleaving a single type of protein at a fairly specific site. An example is collagenase, an enzyme which breaks up collagen, the fibrous protein in connective tissue such as tendon. Collagen also contributes substantially to the 'toughness' of lower quality meat: thus soaking low quality meat in collagenase can tenderize it.

The third use of proteases is in biomedical applications. Many of the biopharmaceuticals planned or in development have protease activity (such as the thrombolytics), but these are not usually considered to be part of the protease industry. However proteases with broad activities also have biomedical applications in such areas as wound debridement (removal of the thick coat of protein material that forms on the surface of wounds which can slow healing and encourage scarring) and as aids to digestion. Proteases can be used either as food supplements or in the preparation of pre digested foods for people in hospital. Here the enzymes have to be of pharmaceutical purity.

The last use of proteases is in biotransformation reactions. Although the normal reaction of a protease is to cut peptides up, if they are used in conditions where there is very little free water (for example in non-aqueous solvents) or if they are used in conditions where the amino acids are freely available but one of the peptides made from them is removed as soon as it is formed, then proteases can be used to make short peptides. Thus the dipeptide **artificial sweetener** Aspartame can be manufactured from a derivative of aspartic acid and methyl phenylalanine using a protease to join them together.

Protein crystallization

A key part of most ways of determining a protein's three-dimensional structure, and hence being able to use that structure to design drugs, is making crystals of the protein. This is difficult, as protein molecules are not as well-behaved as simple salts, and the larger they are the worse behaved they are. The trick is usually to perform the crystallization very slowly and in exactly the right solutions — finding the right solutions can take a lot of expertise and time.

Novel approaches to protein crystallization include crystallizing under high pressure and in space. High pressure reduces the amount of movement in the protein molecule, enabling much faster crystallization in some cases. Crystallizing in free fall means that the crystals do not have to touch the side of the container they are in, and so their growth is not affected by that container. Eight companies and ten research institutes had protein crystallization experiments on the Space Shuttle *Columbia's* mission in January 1990.

The study of the protein crystals so formed is called protein crystallography. It is done with X-rays: the pattern of X-rays that are diffracted from a protein crystal is extremely complex, and depends on the way that all the atoms are arranged inside the crystal. From a good pattern the atom distribution (or more exactly the distribution of electric charge, i.e. the electron density) can be deduced. X-rays can come from a conventional X-ray tube, but a more popular source nowadays is synchrotron radiation, because it is highly monochromatic (i.e. has only one wavelength) and is very intense.

Protein engineering

Protein engineering is the design, production, and analysis of altered, non-natural proteins. This can be a Herculean task if a natural protein is not used as a starting point, so usually protein engineering involves modifying existing proteins.

Protein engineering has a number of aims:

- Improving protein stability. Protease enzymes that have been genetically modified for greater stability are now on the market.

- Altering the substrate specificity of an enzyme. Most enzymes catalyse only a very narrow range of reactions, and it would be helpful to be able to alter that range so that they acted on other, more commercially useful products. Protein engineering can seek to do this by altering the amino acids around the active site of the enzyme, the bit of the molecule which actually binds onto the substrate and catalyses the reaction. By altering the amino acids the forces holding the substrate in place are altered, and hence the molecules which the enzyme best recognizes are altered. A spectacular example of this was the conversion of malate dehydrogenase to lactate dehydrogenase, two enzymes which catalyse similar types of reactions on different substrates. Unfortunately, neither MDH nor LDH are particularly useful enzymes, and this has not been successful on any commercial enzyme.

- Altering pharmacological action. Much protein engineering is aimed at biopharmaceuticals. In this field it seeks to alter the biological activity of proteins which have effects which can be harnessed as drugs, by making the effects more potent, more specific, coupling them to targeting mechanisms so that they only effect a few cells or cell types, improving their survival time in the patient, or reducing side-effects.

See also **Pharmaceuticals proteins, Protein stability**.

Protein sequencing

Determining the sequence of amino acids in a protein is done chemically via a cycle of reactions which chop one amino acid off at a time. There are several machines which perform these quite complex sequences of reactions automatically. The number of amino acids that can be determined depends on the amount of protein available and the nature of the amino acids. None of the reactions in the cycle is 100 per cent efficient, and the efficiency varies to some extent depending on which amino acid is being removed for analysis. Thus after a while, the amount of each amino acid being released by the reaction cycle becomes too small to detect against the 'noise' of other amino acids released from those proteins which were *not* broken in previous cycles. Clearly, also, the protein must be reasonably pure, as otherwise the result is a mixture of amino acids at each step.

The standard chemical method is called the Edman degradation. The process starts from the amino end of the protein (the N-terminus). In some proteins the N-terminal amino acid has another small chemical group attached to it — usually a methyl, acetyl, or formyl group. The presence of this group makes it impossible to start the reaction cycle. Then some pre-preparation of the protein is needed before a sequence can be determined.

Other methods including the use of mass spectrometry (MS), especially fast atom bombardment (FAB) mass spectrometry, are gaining popularity. Short peptides can be sequenced in one experiment using tandem FAB-MS. This is mass spectrometry where there are two MS machines hooked together, one to fragment the protein into bits and separate the bits, the other to analyse the bits. MS methods can cope with blocked peptides, as well as with glycoproteins, lipoproteins, and proteins which have been chemically altered in other ways. However it is relatively insensitive, needing milligrams of pure protein to work successfully.

Because of the difficulties in sequencing proteins, and the limit of around 40 amino acids from any one peptide which can be sequenced in one experiment, many workers prefer to clone the gene for the protein (if they can) and sequence the DNA, using the genetic code to deduce the amino acid sequence of the protein. There are potential problems with this approach, however (*see* **Genetic code and protein synthesis**).

Protein stability

Proteins are not very stable in chemical terms: they are easily denatured (i.e. converted to inactive forms) by heat, acids, alkali, and by some chemicals such as urea and guanidine which are known as chaotropic agents. Denaturation occurs when the protein chain of amino acids, usually folded into a specific, tight coiled conformation, unfolds: the carefully arranged three-dimensional structure of its surface is lost, and usually whatever its function was is lost with it. Chaotropic agents are thus called because they induce this 'chaotic' transition in proteins.

If enzyme reactions could be carried out at higher temperatures, or antibodies made more stable so that they could last longer, biotechnologists would be very pleased. So there is a lot of work in trying to improve protein stability. The lines of work are:

- Using another more stable enzyme, especially from a thermophilic bacterium.

- Increasing the number of disulphide bonds within the protein. These bonds, formed between cysteine residues in the protein once it has folded up into its proper shape, help to lock it into that shape.

- Increase internal hydrophobicity. Often the amino acids that end up inside the correctly folded protein are water-hating (hydrophobic) amino acids: if the protein is unfolded, they are exposed to water, a process which requires energy and which therefore tends not to happen.

- Add other stabilizing interactions. A wide range of other interactions between amino acids help to hold a protein in its correct state. These include hydrogen bonds and ion (or salt) bridges.

In all these latter three cases, the protein engineer aims to add or alter amino acids to increase the number of stabilizing interactions in the protein. This needs a detailed understanding of the three-dimensional structure of the protein, information which can be very difficult to obtain.

Proteins can also be stabilized by adding specific stabilizing agents to their preparation. Very few enzymes are sold as pure protein — most have many other materials in their 'formulation' to stabilize them. Some of these can have a dramatic effect, extending lifetimes from hours to weeks. Exactly what is in each stabilizer depends on the enzyme concerned.

Folding and stability is also important when a protein is to be made by **recombinant DNA technology**. Frequently a protein made at high levels in a bacterium is not made in its native (i.e. normal) conformation. This may be because the protein precipitates inside the cell as an inclusion body, or it may be because the protein is synthesized or modified in different ways in a bacterial cell. Thus part of the purification procedures for many recombinant proteins involve steps which partially unfold the protein and then refold it again, this time under conditions which allow it to fold up properly. (This can also help purification, by selectively unfolding and refolding the desired product: contaminant proteins fail to unfold, or fail to fold up again, and so can be distinguished from the product.) Clearly, it must be relatively easy to fold up the protein if this strategy is to work — some proteins cannot be refolded into their native structure once they have been unfolded.

See also **Disulphide bond, Hydrophobicity, Protein crystallization, Thermophile**.

Protoplasts

Many cells are surrounded by a tough, thick cell wall. Plant, fungal, and most bacterial cells have cell walls. A protoplast is such a cell from which the cell wall has been removed, leaving the cell naked and surrounded only by its plasma membrane.

There are a number of reasons for wanting to do this, but all involve the strength of the cell wall itself. Often plant breeders wish to fuse the cells of two quite different plants which cannot be cross-bred by conventional methods. However the cell wall gets in the way. Again, getting DNA into plant or yeast cells for genetic engineering is extremely difficult, with the cell wall essentially impervious to any large molecules. (Getting DNA into bacteria is an exception because bacteria have mechanisms for absorbing DNA from the medium surrounding them.) Thus many manipulations using these types of cells require you to start with protoplasts.

Plant and yeast protoplasts are generated by dissolving their cell walls with appropriate enzymes, which will digest the carbohydrate (plant) and chitin (yeast) in the cell wall without affecting the lipid-and-protein cell membrane.

The cells of yeasts and some plants can be regenerated from protoplasts, providing that the cells have not been shocked too much when they were turned into protoplasts in the first place. Thus protoplasts which have been genetically manipulated can be turned back into normal cells. This is desirable because protoplasts are very fragile — even more fragile to chemical and physical attack than animal cells in culture — so they are extremely difficult to use in a commercial biotechnological process. Plant cells which have been regenerated in this way can then be used to regenerate whole plants, so using protoplasts of plant cells can be a step in plant genetic engineering.

Purification methods: large scale

One of the central parts of the downstream processing of a fermentation product is purification. Large-scale purification methods are used to take a crude fermentation supernatant or cell homogenate and isolate the product from it in a fairly pure form. Industrial enzymes are often sold in this semi-pure form as a bulk product; if they need to be really pure, then they have to go through a second purification step, often on a small scale. Purification of the cells from a culture is usually called harvesting, and relies on rather different methods.

There are a range of purification methods which are cheap enough to use on large volumes of material including the following.

Salt precipitation. Adding salt so that a particular group of proteins precipitates from solution. Simply adding water to the precipitate usually makes them dissolve again.

Liquid–liquid separation. Also called two-phase separation, this uses the fact that the material that is wanted will dissolve well in one solvent while

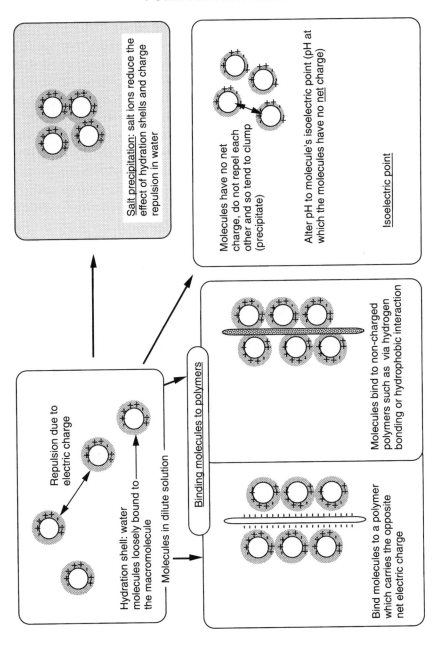

Salt precipitation: salt ions reduce the effect of hydration shells and charge repulsion in water

Molecules have no net charge, do not repel each other and so tend to clump (precipitate)

Alter pH to molecule's isoelectric point (pH at which the molecules have no net charge)

Isoelectric point

Repulsion due to electric charge

Hydration shell: water molecules loosely bound to the macromolecule

Molecules in dilute solution

Binding molecules to polymers

Molecules bind to non-charged polymers such as via hydrogen bonding or hydrophobic interaction

Bind molecules to a polymer which carries the opposite net electric charge most of the impurities will not. The two are intimately mixed, and then separated (by allowing to stand, by filtration systems, or by gentle centrifugation). This only works, of course, if the two liquids are immiscible. This can be performed several times, reducing the amount of contaminant in the sample phase each time. For large-scale preparations it is essential that the two phases are cheap, as it rare that they can be recycled efficiently. One is usually water (as that is the basis of the culture medium) and so the other is a material like benzene, ether, or oil.

Two-phase aqueous extraction. Here the protein is shaken up with a polymer-based mixture which, on leaving to stand, separates into two distinct layers (polyethylene glycol (PEG) and salt will do this trick, for example). The conditions are arranged so that the product ends up in one layer and most of the contaminants in the other.

Polymer precipitation. Some polymers, particularly PEG (polyethylene glycol) can bind gently to proteins and make them precipitate of their own accord.

Heat denaturation. This is simple and effective if the protein is heat stable (thermostable): the mixture is simply heated up and most of the proteins denature, and so 'coagulate' and settle out of solution. The thermostate protein remains soluble. This only works with some proteins. It can also be used in some conditions to separate proteins from non-protein products (e.g. metabolites).

Isoelectric point separations. Most proteins are quite insoluble at a particular pH (their isoelectric point or pK_i). If acid or alkali is added until the pH (the 'acidity') of the solution is at this isoelectric point, then those proteins will precipitate. Again, adding water usually redissolves the precipitate.

See also **Harvesting, Purification methods: small scale.**

Purification methods: small scale

As many biotechnological products have to be extremely pure, for use as drugs or in producing fine chemicals, the relatively crude purification methods which isolate them from large-scale culture are not good enough. A further step of purification is needed. There are many such methods, but relatively few are used commercially. Most are chromatography methods. Here the mixture is passed down a tube which is packed with some material to which some components in the mixture stick and others do not. It does not matter whether the product you want sticks or not, providing the contaminants do the opposite.

- Affinity chromatography (*see* **Affinity chromatography**).

- Gel filtration. This is a chromatography method in which the molecules are separated by size.

- Ion exchange. This separates molecules according to their charge. As the charge of a molecule depends on the pH, a combination of varying pH and ion exchange chromatography can prove very effective at purifying proteins.

- Hydrophobic chromatography. This type of chromatography uses the different affinity that different molecules have for hydrophobic materials, i.e. for materials which are 'water-haters' like plastics (as opposed to hydrophilic — 'water lover' — materials like paper).

Popular versions of all chromatographic separation methods are FPLC and HPLC, which have been scaled up from laboratory tools to production methods in some cases. HPLC — high pressure liquid chromatography — pumps the mixture through the chromatography 'column' at very high pressures, ensuring very precise separation in a short time. FPLC- *M* fast protein liquid chromatography — is a more specialized technique for separating proteins, which, because many biotechnological products are proteins, has found wide use. The pressures used in FPLC are much lower than in HPLC, so the apparatus can be substantially cheaper.

See also **Chromatography**.

Rational drug design

This is a very rapidly growing area of biological effort, in part because it offers an alternative to the exhaustive screening programmes by which drugs used to be discovered and in part because it is done on a computer and produces coloured pictures. The basic technology is to model the molecular structure of the target of a drug, and then design a drug molecule which will fit it. This contrasts to the alternative, which is to screen a large number of compounds for drug activity, choose the most promising and make a whole lot of variants, choose the most promising of them and repeat until a suitable drug is found.

Rational drug design involves knowing the chemical structure of the drug's target which almost invariably means knowing the structure of a protein. Protein structures are very difficult to obtain: it is relatively easy to obtain the amino acid sequence of a protein if it can be purified, but determining how the peptide chain folds up in space is difficult. Discovering the structure usually involves cloning the genes for the proteins to which the drugs are to bind, and making them in large amounts in an **expression system**. The protein must then be crystallized, and the structure of the crystals deduced using X-ray analysis. This is a long and difficult process. A potentially more rapid process is to deduce the structure of the protein from the gene sequence: however this is not possible yet unless you already know a lot about it or a related protein.

There are a range of other technologies for directing the search for new drugs, such as receptor binding studies.

See also **Computational chemistry, Protein crystallization, Receptor binding screening.**

Receptor binding screening

This is one of the biotech-based methods for discovering conventional (i.e. 'chemical') drugs. The method relies on the fact that many drugs act by binding to specific proteins (receptors) on or in cells: these proteins usually bind to hormones or to other cells, and control the cell's behaviour, although they may be enzymes or structural elements of the cell. The drug interferes with the normal role of the protein.

Finding a drug which has a particular effect on a cell or animal involves exposing the cell or animal to the drug and then looking for the often subtle effect. Receptor binding assays isolate the receptor protein, and then search for chemicals which latch onto that receptor. The ones that do *may* not be good drugs, but the ones that do *not* are pretty sure *not* to be, so you have narrowed the field.

The problems are twofold. Firstly, you have to know what the relevant receptor is. (Indeed, for many drugs there may not be any receptor which is sufficiently specific, or localized on sufficiently few cells. Anti-cancer drugs suffer from the problem that cancer cells often do not have any

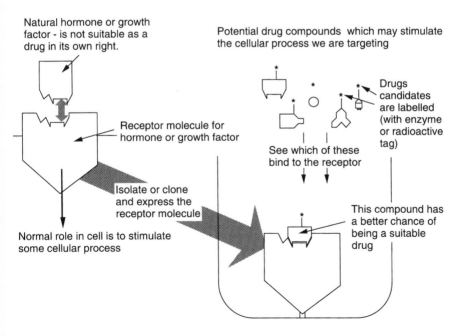

Natural hormone or growth factor - is not suitable as a drug in its own right.

Potential drug compounds which may stimulate the cellular process we are targeting

Receptor molecule for hormone or growth factor

Drugs candidates are labelled (with enzyme or radioactive tag)

See which of these bind to the receptor

Isolate or clone and express the receptor molecule

Normal role in cell is to stimulate some cellular process

This compound has a better chance of being a suitable drug

unique proteins which the drug can 'target'.). Secondly, even when you have identified it, there are usually only a few thousand molecules per cell, so you have to process several kilograms of mouse to get a few milligrams of receptor. So the receptors are often isolated from cloned cell lines which have been selected to over-express them, or from cloned genes which express the receptors in yeast or mammalian cells.

There are several companies involved in using receptor screens, including most of the major pharmaceutical companies and several small companies such as Protos and Receptortech which are dedicated to '**rational drug design**'. The most spectacular is Affymax, a company that has developed chemical methods for depositing huge numbers of peptides and oligonucleotides onto small silicon 'chips', and is using them to screen those peptides and other compounds for their ability to bind to receptors.

Recombination DNA: bits and kits

There are a number of pieces of the technology of DNA cloning that are referred to commonly without further explanation. The more commons enzymes and reagents are:

- Adaptor/linker. These are short oligonucleotides that are used to join disparate DNA molecules together. To do the actual joining, DNA ligase is needed.

- DNA polymerase. An enzyme that makes DNA. To do so, it must have a DNA molecule to copy (the template) and a short DNA molecule to start with (the primer). It then adds bases onto the primer, copying the template until it gets to the end.

- DNA ligase. Also sometimes T4 DNA ligase. This enzyme joins two double-helical DNA molecules together to make one longer one.

- Klenow. A version of DNA polymerase.

- Methylation. This is a process (again carried out by specific enzymes, methylases) which puts methyl groups on to specific bases on DNA.

The presence of these methyl groups can stop some restriction enzymes cutting at that site.

- Restriction enzymes. These are enzymes which cut double-stranded DNA at very specific base sequences, and nowhere else. Thus they cut cloned DNA into a few pieces only. The place at which they cut is called a restriction site, and the map of all such sites on a clone is called a restriction map.

- Reverse transcriptase. An enzyme that makes DNA, but uses an RNA template to do so, not a DNA template.

- RNA Polymerase. There are several of these around, notably SP4 RNA polymerase. These are used to make an RNA copy of a DNA. It needs a template but does not need a primer.

- Taq polymerase. Another DNA polymerase made from *Thermus aequaticus*, and an enzyme that is stable when heated to 95°C.

There are a lot of 'kits' on the market, collections of reagents, enzymes, DNAs, and even organisms that have been developed as packages that work together to process the purchaser's samples. Among the more common are packaging kits (which are used in bacteriophage cloning), *in vitro* transcription and translation kits (which carry out transcription and translation in a 'test-tube'), kits for **site-directed mutagenesis**, kits for labelling DNA with radioactive, fluorescent, or chemical labels and so on. There is a school of thought that there are so many kits around that molecular biology has been reduced to a game of putting together the right kits and writing up the result. Having done it both with and without kits, I think that kits have a lot going for them in allowing the scientist to concentrate on doing creative experiments rather than making all the reagents he needs.

Recombinant DNA technology

This is the blanket term for the technologies which have made the recent boom in biotechnology possible. It is also called biomolecular engineering, especially in France (ingenieur biomoleculaire). Recombinant DNA techniques allow the biotechnologist to isolate and amplify a single gene from all the genes in an organism, so that it can be studied, altered and put into another organism. The technique is also known as gene cloning (because you produce a whole lot of genetically identical genes), and the result is sometimes called a gene clone, or simply a clone. An organism manipulated using recombinant DNA techniques is called a genetically manipulated organism (GMO).

Recombinant DNA technology covers the following areas:

- Isolating genes. This involves splicing the gene of interest with a vector and putting the result into a suitable organism, usually a bacterium or yeast. This new DNA is made of at least two bits of DNA (the target gene and the vector) and is called a recombinant DNA. The combination then grows, multiplying up the gene–vector combination as it does so, to produce a clone of cells. The DNA is said to have been 'cloned into' the vector.

- Identifying and characterizing genes. This involves finding which clone contains the gene you want. This is done using biochemical methods of increasing power to discriminate one gene from another, culminating in DNA sequencing (*see* **DNA sequencing**).

- A related technique is sub-cloning. This takes a large gene clone and breaks it up into smaller pieces, making a new clone from each. This means that what was originally a large piece of DNA is now in smaller, more convenient pieces. This is often used to take a large piece of DNA with many genes on it and separate the genes into one-gene-clones.

- Modifying genes. This involves replacing anything from a single base to a whole chunk of the gene with other DNA, using (*inter alia*) site-directed mutagenesis.

- Putting the genes into another organism. In some cases this is not necessary, as it is knowledge about the gene that is required. However for biotechnologists, putting the gene to use is usually essential, and so the gene is put into another organism using one of the following methods: transfection, transduction, transformation, biolistics, electroporation, or microinjection.

See also **Biolistics, Electroporation, Homologous recombination, PCR, Site-directed mutagenesis, Transfection, transduction, transformation vector**.

Regulation

Biotechnologists sometimes complain that the industry is heavily over-regulated, but in practice it is no more regulated than many other industries, and especially those which rely on relatively new technology. Several aspects to the regulation of biotechnology are covered in this book.

- Patents and intellectual property rights.
- Safety of the microorganisms and genetically engineered constructs.
- Safety of genetically engineered organisms to be released into the outside world.

See also **Microorganism safety classification, Patents, Regulation of organism release**.

Regulation of organism release

Regulations about the deliberate release of organisms, and particularly genetically manipulated organisms, vary widely. The US has a fairly consistent set of regulations controlled by the Environmental Protection Agency (EPA). European regulations vary enormously from the extremely restrictive (Denmark) to the extremely liberal (Italy, Greece) as judged by the American yardstick. By the end of 1989 there had been 140 deliberate release experiments in the US, about half that number in Europe. Deliberate release experiments in the US are the subject of intense and often very public debate about their safety. In Europe, where public access to private data is more limited, laws such as Britain's Environmental Protection Law allow for public access to private data about potential deliberate release experiments so as to allow for the same level of public involvement in deliberate release experiments as the US experience has taught Europeans to expect. By the end of 1992 all countries in the European Community will have to abide by European directive 91/220 regarding notification and control of deliberate release.

Regulatory authorities (US)

There is a wealth of regulatory bodies in the US which oversee the biotechnology industry. In general terms, their requirements for the safety and efficiency of biotechnology products are the most stringent, so most biotechnology companies aim to satisfy US regulations on the assumption that, as well as being the largest single market for almost all biotechnology products, it is also the hardest to get into.

Some important regulatory agencies are:

- National Biotechnology Policy Board (NBPB). Provides an advisory scientific board for the Secretary of Health and Human Services on the scientific issues behind biotechnology regulation.

- President's Office of Science and Technology Policy (OSTP), which replaced the previous Biotechnology Science Co-ordinating Committee (BSCC). It has a broad remit to evaluate the scientific basis of biotechnology regulation, and advise the federal government on regulatory issues. The board's remit and membership overlaps substantially with the NBPB.

- FDA. Food and Drug Administration. Oversees and regulates all medical drugs and devices, and new food and cosmetic products, assuring that they work, and are not harmful. An autonomous agency, it is the principle regulatory agency which any company must appease in order to launch a new drug or medical device onto the market. In general, FDA regulations set the pace for other countries in biotechnology, because the US market dominates biotechnology products and so all companies want to make sure that their products and processes fit FDA regulations. FDA regulations cover the efficacy of a drug (and hence how its trials are done), how it is manufactured (*see* **GLP/GMP**) and how it is formulated. It is notable that, since 1958, the burden of proof that a drug or food additive is safe falls on the producer, and the FDA is not responsible for proving that it is *not* safe.

- EPA. The Environmental Protection Agency. Has responsibility over deliberate release of organisms into the environment.

- Healthcare Financing Administration. Developing a biopharmaceutical is time-consuming and expensive, and the number of patients which can benefit from it are usually small compared to many conventional drugs. The HCFA has a substantial role in determining the acceptable price for a new drug, and hence whether the company which developed the drug can make enough to recoup its investment and generate funds for future research. This has affected biopharmaceuticals particularly: Streptokinase, an established 'clot-buster' drug, costs US$186/dose. tPA, a genetically engineered alternative which some studies say is no more effective, costs US$2200/dose. The HCFA's comments are particularly relevant as most biopharmaceuticals — indeed most drugs — are aimed at the elderly, many of whom are covered by the federal Medicare programme (which has 34 million elderly and disabled clients) in the US.

Replica plate

This is a simple technique for reproducing and selecting bacteria. A number of bacteria are grown on a petri dish. A pad (traditionally sterilized felt) is lowered careful onto the plate, and when it is taken off some of the bacteria stick to it. It is then lowered onto another plate, where some of the bacteria stick. This second plate then carries a replica of the pattern of organisms on the first plate. The replica plate may now be incubated, and the bacteria on it tested quite destructively for some property. The ones that come out as having the best result are then identified, and the corresponding group of organisms on the original plate can be identified because they are in the equivalent place.

REPLICA PLATE

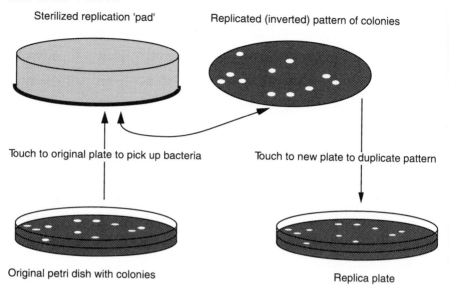

Sterilized replication 'pad'

Replicated (inverted) pattern of colonies

Touch to original plate to pick up bacteria

Touch to new plate to duplicate pattern

Original petri dish with colonies

Replica plate

Related techniques are the 'plaque lift' and 'colony blot' methods. In these cases the 'pad' is a special filter membrane; which is lowered onto the plate. After some microorganisms have stuck to the filter, it is removed and treated to break the cells and liberate the DNA and proteins that were inside them. Specific biochemical tests then detect whether the DNA or protein we want is among them. Again, the bacteria or bacterio-phage that contained those proteins or genes can be identified by their position on the original plate.

Retroviruses

Retroviruses are viruses whose RNA genes are copied onto DNA as part of their life cycle. The DNA is then usually inserted into the DNA of their host cell, and it can remain there for many cell divisions as a 'provirus' until some signal triggers it to be transcribed onto RNA and so be translated into viral proteins, making more virus. The only thing distin-guishing the provirus from any other DNA in the cell is its base sequence.

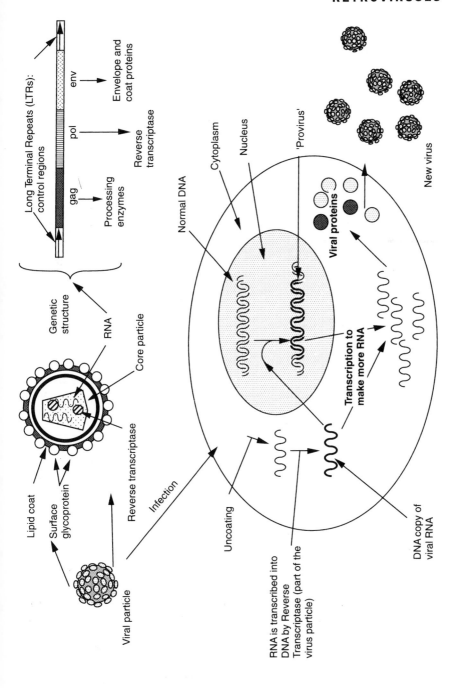

Genetic structure

Long Terminal Repeats (LTRs): control regions

gag → Processing enzymes

pol → Reverse transcriptase

env → Envelope and coat proteins

RNA

Core particle

Lipid coat

Surface glycoprotein

Reverse transcriptase

Viral particle

Infection

Normal DNA

Cytoplasm

Nucleus

'Provirus'

Viral proteins

New virus

Uncoating

RNA is transcribed into DNA by Reverse Transcriptase (part of the virus particle)

Transcription to make more RNA

DNA copy of viral RNA

283

RETROVIRUSES

Retroviruses are of interest to biotechnology for two reasons. Several retroviruses are of medical importance. The AIDS virus HIV is a retrovirus, as are several other immune-system targeting viruses (the HTLV family), and some viruses which can cause cancer in laboratory models (the oncogenic retroviruses). Thus retrovirus biology is very important to the search for treatments and cures for AIDS.

The ability of retroviruses to infect a cell and then insert their DNA copies into the chromosomes of that cell, has also been harnessed to make DNA cloning vectors which can get foreign DNA reliably integrated into mammalian chromosomes. These have been used to transfect mammalian cells, and to create transgenic animals by infecting embryonic carcinoma (EC) cells with retroviral vectors. The vectors have to have only part of the viral DNA in them, because otherwise they would simply produce infectious virus. Thus the ideal retrovirus-based vector has those genes which are needed to insert the DNA into the chromosomes but no others. Sometimes this requires that the engineered vector is infected into the cell along with a 'helper virus' which provides some of the genetic functions necessary but which does not itself get into the cells.

Retroviruses are a specific category of a class of genetic elements called retrotransposons, genetic elements which can copy themselves into new places in the genome through an RNA intermediate. Many genetic elements which are valuable to plant geneticists are retrotransposons: they are used for shuttling genes around plant chromosomes or for causing selective mutations in plants.

See also **AIDS, Chimaeria, Transgenic animals: applications**.

Reverse genetics

Reverse genetics is the type of genetic analysis which starts with a piece of DNA and proceeds to work out what that does. By contrast normal genetics ('forward genetics') starts with the phenotype — what the organism looks like — and proceeds to work out what the genetic structure is, ultimately decoding the DNA itself.

Such feats of gene cloning as the isolation and characterization of the cystic fibrosis gene are often called reverse genetics: however, although these use an impressive panoply of recombinant DNA techniques, they still start with an observed phenotype (the disease) and work via ever more detailed genetic techniques to a genetic explanation of what is going on. Reverse genetics has been used, for example, in understanding the genetic structure of a range of viruses, including the AIDS virus. Here the DNA structure is known in detail, but what it does is not known. So mutations are found or made in the DNA, and then their effect on the phenotype is discovered. In this way the function of those bits of the gene is worked out.

Reversed phase biocatalysis

Some enzymes work on reactants or products which are fairly, or almost completely, insoluble in water. Others work using water as a substrate, and it would be useful to be able to remove the water from the reaction to make it 'run backwards'. In both cases, it is useful to be able to operate an enzyme reaction in a solvent other than water.

Organic phase catalysis and supercritical fluid catalysis offer ways of doing this (*see* **Organic phase catalysis, Supercritical fluid enzymology**), but an alternative which is not so radical is reversed phase biocatalysis (also biphasic biocatalysis), in which an enzyme dissolved in microscopic droplets of water is suspended in an organic solvent containing the reaction substrate and/or product. The enzyme's substrate diffuses out of the solvent in very small amounts, is acted on by the enzyme, and diffuses back into the solvent. Because the droplets are very small, the rate of diffusion is very fast and so the reaction proceeds at a useful rate.

A variation is to use a solid support to hold the enzyme in a totally organic solvent. The solid support has a single-molecule layer of water adsorbed onto its surface: the enzyme sticks to that, and is simultaneously immobilized (so that it is easy to remove as part of the particulate solid once the reaction is done), activated by the water, and stabilized by immobilization. Inorganic materials such as silica or celite are usually used.

These systems have the advantage that you do not have to dehydrate the enzyme so thoroughly before the reaction (organic phase catalysis needs a thoroughly dehydrated enzyme to work properly), and so can be much easier to get working.

RFLP

This abbreviation, standing for restriction fragment length polymorphism and sometimes pronounced 'riflip', is in common use in a range of applications of DNA technology to genetics. It means a piece of DNA which varies between two individuals. It is irrelevant whether the DNA has a function or not, or whether the variation is important. The term refers only to the way of detecting the variant, which is by the use of the very specific DNA-cutting enzymes called restriction enzymes. The essence of an RFLP is that one variant is cut by a particular enzyme at one site, the other is not. This means that the fragments produced by that enzyme on those DNAs have different lengths.

RFLPs have found wide use as 'marker' genes for genetic studies. Here

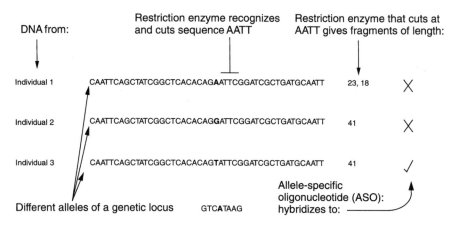

the RFLP is used to detect when a piece of DNA has been inherited by an individual from one parent (rather than the other). If the RFLP is near to a gene which we wish to track but cannot detect directly, then there is a fair chance that the target gene has been inherited along with the RFLP. The RFLP is termed a linked marker, as it is physically and genetically linked to the gene in which we are actually interested.

A related term is the allele-specific oligonucleotide (ASO). This is an oligonucleotide which will hybridize to the DNA from one individual but not to that of another, because the DNAs differ by one or two bases. The variant forms of the DNA are called alleles. Both RFLPs and ASOs have found substantial use in human genetics, and in animal and plant breeding programmes.

Ribozymes

Also called catalytic RNA, these are RNA molecules which catalyse chemical reactions, often the breakdown of other RNAs. Their discovery in the mid-1980s overturned the idea that only proteins could be biological catalysts, and won Cech and Altman the Nobel prize.

Ribozymes have potential in two areas. They are widely touted as potential pharmaceutical agents, as their action against other RNAs can be extremely specific. They could, for example, attack a viral RNA without affecting the normal RNAs in a cell. Thus they have potential as antiviral agents and, through their potential ability to attack RNAs from oncogenes, as anti-cancer agents. Ribozymes as therapeutics are still a research technology, however. Although very specific in the test-tube, like antisense RNA they can have unexpected effects when introduced into cells. There is also the problem of how to get them into cells. RNA is destroyed very easily by chemical or enzymic attack, and so has to be protected by encapsulating it in (for example) liposomes to get it to the cells it must affect.

The other area is to use ribozymes as industrial catalysts, selecting suitable catalytic activities through Darwinian cloning.

See also **Antisense, Darwinian cloning**.

Scale-up

Scale-up is the process of taking a biotechnological production from a laboratory scale to a scale at which it is commercially useful. A few biotechnological processes can be run on laboratory scale systems (for example, the production of reagents for research use, such as monoclonal antibodies). All others have to be done on much larger installations than a research laboratory can handle.

The difficulty with scale-up is that a tonne of fermenting bacteria seldom behaves in the same way as a gram of the same bacteria, unless it is divided into a million separate tubes. In general, it is not possible to take the conditions which have worked well in the laboratory and apply them to an industrial process. Instead, they are gradually adapted to ever larger scales of production, each step usually being something like a ten-fold increase in size over the previous one. At each stage, the optimum amount of various metabolites, and mechanical parameters (such as stirring rate, air-supply method and rate) must be determined, based on the biotechnologist's experience with previous production systems and a general knowledge of scale-up procedures. There is some mathematical modelling available to help this, but even so experimentation is essential.

The problems of scale-up were not well understood by the early genetic engineers, and so in the mid-1980s there was a severe shortage of scientists skilled in this field. However it is now understood that a marvellous laboratory result does not automatically translate into money in the bank, as scale-up might be prohibitively expensive.

Scanning tunnelling microscopy (STM)

This new type of microscope has been promised to be the ultimate way of discovering the structure of biomolecules (among other things). A related technique is the atomic force microscope. In essence, an ultra-sharp needle point is scanned slowly over something, and either the force on the needle or the electric potential of the needle tip is monitored. When the tip encounters an atom sticking out above the general surface, the extra force/current is measured. By scanning back and forth across a surface, a picture of the hills and valleys can be built up on an atomic scale.

There are two application areas in biotechnology, neither advanced beyond a laboratory curiosity stage. The first is directly detecting the physical shape of complex molecules, getting around the need for the pure crystals that X-ray methods need. Arscott and Bloomfield at Minnesota University have produced pictures of the double helical structure of a synthetic DNA using STM. By hitting molecules under the STM with light (and so altering their shape), something can be deduced about the chemical nature of individual bits of a new molecule as well as their size and shape.

The other, even more radical idea is to use STM as a way of actually moving atoms around, creating new chemical entities. So far this has been confined to drawing letters with individual atoms on crystal surfaces, the atoms being xenon (at IBM in San Jose) or sulphur (at Hitachi in Tokyo). In principle this could lead to the direct fabrication of new biomolecules which would be enormously hard to make by any conventional method: however, this is definitely 'Buck Rogers' stuff at the moment.

See also **Molecular computing**.

SCP (single cell protein)

Coined in 1966 at Massachusetts Institute of Technology (MIT), the term single cell protein refers to protein biomass used as a food additive for animals or people. Either isolated protein or whole bacterial cells (suitably processed) may be called SCP.

The drive to develop SCP came from the realization that the 'food shortages' seen in many third world famines were primarily shortages of protein, not of food bulk *per se*. Similarly, the limiting factor in many animal feeding systems is how much protein is available for animal growth, not the total calorific content. The idea behind SCP technology was to use bacteria, growing on a cheap carbon substrate and with a cheap nitrogen source such as ammonia, to make protein fit for human, or at least animal, consumption.

As with many large-scale fermentations, the key to making SCP economic is to find a carbon source cheap enough. Oil and natural gas have been tried, but are only marginally economic even when the oil and gas prices are low. Methanol made from natural gas is a good potential substrate as bacteria find it easier to use (they need less oxygen to grow on methanol than on methane, and it is very soluble in water). ICI developed a large-scale biomass process based on a methanol-using bacterium (*Methanococcus*) to produce a partially purified protein product ('Pruteen'). The production scale plant had a volume of 1000 m^3 and a capacity of 70 000 t of SCP per annum: despite the economies of scale, however, this was at best only marginally economic, despite ICI's use of genetic engineering to improve the effectiveness of the bacterium's metabolism in using ammonia to make the protein.

The problems with SCP are that most microorganisms have a much higher nucleic acid (DNA and RNA) content than animals or plants, which can cause health problems, that microbial cells can absorb or make toxic materials during fermentation, and that the cells themselves may be extremely indigestible or allergenic. This has limited the use of SCP in human food, and meant that most effort has gone into using it as an animal feed supplement. In this use it competes directly with soybean meal and fish meal.

Cellulose, wood, waste starch, paper processing effluent, and other complex sources of carbon have all been suggested as potential SCP substrates: however none of these are efficient enough to be economic.

Sea water

There have been many and varied plans for extracting metals from sea water, often lured on by the fact that a cubic mile of sea water contains more than 1000 t of gold. However the gold is spread over a very large volume, and no process yet devised is cheap enough to get the gold — or anything else — other than salt and a few other chemicals out of it.

Biosorption and bioaccumulation are biotechnological routes to obtaining value from sea water. The idea is to use bacterial cells to accumulate a specific metal from the water: all you have to do is pass the water over the cells, and then afterwards 'wring them out' into a much smaller volume, resulting in concentrated gold solution. Despite this attractive sounding proposition, it is never economic to do this when the actual costs are worked out, including (for example) the cost of pumping 4 billion tonnes of sea water through your extraction apparatus, and replacing components of the extraction plant regularly as they are eroded by the salt water.

See also **Bioaccumulation, Biosorption**.

Secondary metabolites

Primary metabolites are the chemicals commonly found in most living things, and which are essential for them to live. Compounds such as glucose or glycine would fit into this category. Secondary metabolites are chemicals which are usually unique to one organism or class of organisms, and which are not essential for cell survival. They perform more specialist functions like being involved in specific stages in the organism's life cycle, degrading unusual food sources or (usually) fighting off other organisms. Many of the chemicals that plants or microorganisms produce which are of biochemical interest, including antibiotics, are secondary metabolites.

SECONDARY METABOLITES

Unlike primary metabolites, which most organisms contain most of the time, the production of secondary metabolites is very dependent on the environment of the organism. Thus small changes in culture conditions of an actinomycete (actinomycetes are the most commonly used sources of new secondary metabolites) will dramatically alter how much of a particular chemical they produce.

Plants often produce secondary metabolites as defences against infection or being eaten: caffeine in coffee beans, atropine in foxglove, and the vinca alkaloids in madagascan periwinkle are examples of quite poisonous compounds made to ward off attack. These secondary metabolites are usually not produced efficiently in isolated, cultured cells. However their production can sometimes be stimulated by elicitor compounds or preparations, which are often fungal or plant extracts.

Secondary metabolites are used for many purposes. The two most common are:

- As drugs. Many drugs were discovered when a plant or fungal extract was found to have pharmacological activity. Almost invariably this activity is due to a secondary metabolite. Often the chemical structure of the metabolite is so complicated that it is still extracted from its natural source, as making it by chemical synthesis would be too expensive. Antibiotics are often secondary metabolites, as are alkaloids.

- Flavours and fragrance compounds, other than sweet flavours and salt, are usually secondary metabolites, often from plants. (Meat flavours arise rather differently, from chemical reactions between fats, protein breakdown products, and sugars in the meat.) Several companies such as Universal Foods and Universal Flavours and Fragrances are working on using plant cell culture and cloning methods to produce flavour or fragrance chemicals by fermentation.

Metabolism is usually divided into anabolic pathways — pathways that build up molecules for use by the organism (e.g. those pathways that make amino acids), and catabolic pathways — ones that break down molecules, either for energy or simply to get rid of undesired materials (e.g. breaking down hydrocarbons to obtain energy). Some pathways, especially those at the centre of metabolism (e.g. those that break down glucose) perform both functions, and are called amphibolic. In general, secondary metabolites are the product of specialist anabolic pathways.

See also **Antibiotics**.

Secretion

Secretion is the active export of a material from a cell or organism. The secretion of proteins by bacterial and mammalian cells is very important in protein production by biotechnology. If a foreign protein being produced by a cell can be secreted, then it is usually much easier to purify from all the other proteins that the cell is making, as these mostly stay inside the cell.

Proteins which are to be secreted from a cell have a short peptide on their front end — the signal peptide — which acts as an export label. The signal peptide is chopped off the protein as it is exported (during a step called 'processing'), so the final protein does·not have this extra peptide on it. Genes for naturally secreted proteins code for this peptide. Genes for proteins that are not normally secreted do not, and so this signal peptide must be engineered onto the front end of the 'new' gene. Secretion vectors are expression vectors that allow this. They have a promoter and then a short section of a gene which codes for this signal peptide. Any gene spliced in exactly next to the signal peptide gene will produce a fusion protein — one with the signal peptide joined onto the front of the protein — which should then be exported from the cell.

See also **Expression systems**.

Sewage treatment

Sewage treatment is one of the most widespread biotechnological processes in urban Western societies, which produce huge amounts of human and animal waste. Sewage treatment methods vary widely, but all have a biological basis to break down the organic material in sewage and convert it into something that can be safely discharged into rivers or seas.

SEWAGE TREATMENT

All sewage treatment falls into several stages:

- Filtering — to remove large solid objects (paper, sticks, sand, etc.)

- Settling — to allow the particulate material to settle out. This sludge is usually then composted to decompose any organic matter, and then used for landfill or as fertilizer.

- Biological treatment — the resulting liquid is treated using micro-organisms to remove the remaining organic matter. This treatment can be by

  — trickling bed system, in which the liquid is pumped over mineral or plastic beds with a film of organisms growing on them.

  — activated sludge process, in which the sewage is incubated with organisms derived from the sewage sludge, with air or oxygen pumped through the mixture.

- Further settling — the microbial biomass produced during biological treatment is allowed to settle out, resulting in reasonably clean water. The sludge is either recycled to the fermentation system or further incubated to make fertilizer.

An important aspect of sewage processing is the reduction of the number of organic carbon compounds in the sewage, expressed as biological oxygen demand (BOD). BOD is the amount of oxygen that is needed for microorganisms to use up all the nutrient sources in the water. A lot of organic material in the water implies that the microorganisms in it will rapidly deplete it of oxygen, making it lethal to fish and probably undrinkably contaminated with bacteria. In conventional sewage the organic material is metabolized by the microorganisms in the treatment plant, ending up as carbon dioxide or as biomass. Alternative methods generate methane (biogas) from this material, but this is not a common usage.

Site-directed mutagenesis

This is the introduction of specific base changes — mutations — into a piece of DNA using recombinant DNA methods. There are several ways of doing this, but all those in common use involve using a synthetic DNA (which has the desired mutation built into it) to replace the equivalent piece of DNA in the original gene. This can be done by copying a new version of the gene from the old version, using an enzyme (usually acting on a single-stranded DNA, such as an m13 clone), or by cutting out the old version of the section of the gene required to be mutated and splicing in a new, mutated version.

The alternative to site-directed mutagenesis is some version of random mutagenesis, where the DNA is mutated at random by chemical treatment and the desired mutant selected out from the mixture of results.

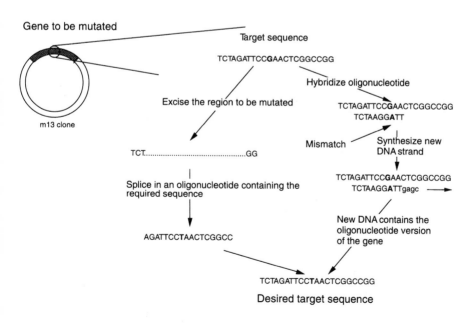

Soil amelioration

The improvement of poor soils, usually using bacteria or fungi. (This contrasts with **bioremediation**, which is the cleaning up of toxins, usually in soils.) Amelioration includes breaking down organic matter, forming humus (i.e. good soil structure), making minerals such as phosphates in the soil available to plants by solubilizing them, fixing nitrogen, and sometimes an element of bioremediation as well.

Soil improvement methods were touted as being the way to 'green the deserts' in the 1960s: however they have not worked all that well, mainly because the deserts are not very promising material to start on for climatic as well as chemical reasons. Most of what was then called soil amelioration is now included in bioremediation.

Solar energy

There has been quite a lot of interest in using biotechnological methods to generate fuels or power from sunlight. This is what plants do all the time, of course, but getting them to do it for man has proven difficult.

The simplest method is to grow plants, and then turn them into fuel: this can be by extremely traditional routes (burning wood) or by growing organisms with high oil contents to make fuel oil. Attempts to make fuel oil using algae have generally proven uneconomic, as has the use of photosynthetic bacteria to make hydrogen. (Bacteria which generate hydrogen or methane from biomass are more successful, and indeed are the basis of biogas technology).

More speculative schemes plan to use the electrochemistry of photosynthesis directly to generate electricity. This can be done either using intact cells (analogous to a bacterial **biosensor**) or by isolating protein complexes from the photosynthetic apparatus and using them as chemical reagents.

Protein complexes considered have included the photosystems (I or II) which transduce light energy into electrochemical potential in the chloroplast, and more specific parts of the photosynthetic apparatus such as the antenna complex which actually captures photons and passes them to a reactive centre. Power outputs to date have been vastly exceeded by the effort and energy needed to make the materials required for the experiment, and the complexity of the photosynthetic apparatus inside a cell makes this an outside chance for a workable system.

An alternative approach is to use a synthetic chemical system. One example is a chemical reaction series based on ruthenium. A ruthenium compound (ruthenium (II) tris (2,2'-bipyridine)) is a reducing agent in its normal state, but a powerful oxidizing agent when excited by blue light. With a metal oxide catalyst and methyl viologen (MV) as an electron acceptor, this compound can transfer electrons from water to MV. The reduced MV can then be used (in theory) to reduce other compounds. Yields are not good enough to make this of more than research interest, however.

See also **Biogas**.

Somaclonal variation

This is the variation seen between the individuals in a clone, and particularly in plant clones. If you separate a plant into its component cells and culture them in the right conditions, then you can get each cell can grow into a new plant. In theory each of these new plants should be genetically identical to their 'parent'. In practice the cell callus — the undifferentiated mass of cells that grows from an isolated cell in culture — can be quite genetically unstable. Cells can double their chromosome complement, lose genes, or even whole chromosomes. When the callus is made to grow a new plant, the plant inherits these genetic variations, and so is not identical to its parent. This variation is somaclonal variation.

This can be a problem or an opportunity for the plant breeder. It is a problem if you want to use plant cloning technology to grow a lot of your price plant: the offspring of most cloning methods will not be the same as the parents. Somaclonal variation has been the bane of potato breeders (the potato is unusually prone to somaclonal variation) and caused severe problems in Unilever's attempts to use micropropagation methods to grow oil palm trees in South East Asia in the mid-1980s. However, it does provide the opportunity to generate new plant types which would be harder or impossible to generate by conventional plant breeding.

Sports and biotechnology

Despite the fact that recreation, and specifically sports, is big business and approaches the chemical and agricultural industries in size, biotechnology has consistently ignored the lighter side of life in preference to healthcare or process industry products. The only major exceptions seem to be discussions of the potential abuse of biotechnology products for sporting advantage. Two specific cases are widely discussed: these may or may not be actual rather than potential abuses, as 'authoritative' rumours abound with little factual support for them.

Growth hormone. The market for growth hormone for medical treatment is quite small: however the market perceived for the drug is substantially larger, and must include some 'indications' not thought to absolutely require growth hormone treatment when the protein was first expressed in bacteria. Two areas of 'new' application are for short stature and for sport. Kabi Pharmacia placed advertisements in the medical literature in late 1991 suggesting that growth hormone could be a 'cure' for the childhood 'condition' of being short (not pathologically short, but just in the lower few per cent of the normal human range for children of that age.) This could be defensible on psychological grounds. An application which is indefensible on medical grounds is the use of growth hormone to try to make people unusually tall, so that they have an advantage in some sports such as basketball. For this to work, growth hormone would have to be administered during early adolescence. The abuse of growth hormone by adults who try to use it increase muscle mass is fairly well established. Rumours that people have tried to acquire growth hormone to give to their children are widespread — whether this is an urban myth, along with the women who put poodles in microwaves and the people who discover rats in hamburgers, or is based on any true event is not clear.

Erythropoietin (EPO). This biopharmaceutical was developed to boost blood production in a number of diseases such as anaemia and kidney failure, where patients lack red blood cells, and where other therapies, especially for leukaemias, have depleted the bone marrow causing the patients to develop an iatrogenic anaemia (that is, anaemia caused by the treatment rather than the disease). There is a suggestion that athletes have used, or have tried to use, EPO to boost their levels of red blood cells above the normal level to give their blood greater oxygen-carrying ability. This may give them greater endurance in long-distance running. It would almost certainly be dangerous, increasing the viscosity of the blood and hence the risk of heart attack, stroke, or haemorrhage. Dutch cyclist Johannes Draaijer died of a heart attack at 27 years old in 1990 under suspicion of using EPO.

Standard laboratory equipment

There are a few pieces of kit which all biotechnologists seem to use and to refer to by trade names analogous to 'Hoover' or 'PC'. Among the more common are:

- Multiwell plate. Also 96-well plate or microtitre plate. A postcard-sized plastic dish with 8 rows each containing 12 small round wells. Used extensively in cell culture and molecular biology for doing reactions where you need to perform the same action on up to 96 samples at once. Machines for washing and for detecting colour in 96-well plates automatically are common.

- Gilson. Any type of micropipettor, a device which will measure small volumes (i.e. 1 μl–1 ml) of liquid routinely. Goes with Gilson tips, the replaceable tips that attach to the end of the pipette unit and are the only part in contact with the liquid.

- Eppendorf. A centrifuge the size of a 'mini' hi-fi deck that sits on the bench: also the disposable 1.5 ml plastic tubes that fit inside the centrifuge.

- Universal. A cylindrical tube with screw cap holding about 20 ml and now usually made of plastic.

Stem cell growth factors

These are compounds, almost always proteins, which act to make stem cells grow faster. Stem cells are cells which, while not themselves critical parts of muscle or blood, grow into cells which make those tissues. Thus they are the 'stem' from which the 'leaves' of tissue arise. Stem cells therefore have two roles: to make more stem cells, and to make their differentiated 'progeny' cells.

The best characterized stem cells are in the bone marrow. These stem cells — about 1 in 100 000 bone marrow cells — make the precursors for all the cells in the blood. These stem cells are called 'totipotent' because they can make any of the various types of blood cells. As their offspring develop they become fixed ('determined') into the tract of making one type of cell or another, and then eventually develop the final characteristics of the cells concerned ('differentiate') and are released into the blood. The same sort of thing happens in muscle, in skin, and in developing nerves (including the brain).

Clearly, if the stem cells are to play their role, there must be a balance between how fast they make new stem cells and how fast they turn into their differentiated daughter cells. Too much differentiation and there will not be enough stem cells left for the future. Too much division and you will end up with a cancer. A battery of controls affect how this balance is regulated: aberrations in these controls can lead to cancer. The controls can also be altered artificially to correct disease states.

The most discussed stem cells are the blood (haematopoietic) stem cells. True stem cell factor (SCF) was isolated in 1990, but a range of other factors which affect cells in various stages of determination and differentiation have been found and their corresponding genes cloned, usually with the aim of developing them for drug use.

See also **Growth factors, Oncogenes**.

Sterilization

There are a number of established ways to sterilize equipment and materials for biological use. Clearly, if a microorganism or cultured cell is to be grown, either for research or for production, it is vital that no other microorganism be present to compete with it, possibly killing it and certainly producing unwanted contamination. Thus sterilization is a vital part of any biotechnological process.

There are four generally used approaches.

- Heat. All organisms are susceptible to heat, although some are more susceptible than others. Heat can be dry heat or wet. Wet heating to 121°C in an autoclave (essentially a large pressure cooker) is a very popular way of sterilizing equipment and reagents, as it is cheap and easy to do.

- Chemicals. Many chemicals are inimical to life. Extremely corrosive materials such as chromic acid are used to strip all biological residues off glassware. However more benign biocides — chemicals which kill microorganisms but leave most other things intact — are more commonly used. Many are used as cleaning agents, as, unless swallowed, they are relatively harmless to humans. A variation on chemical treatment is treatment with biocidal gas, usually ethylene oxide. This has the advantage that equipment being sterilized does not then have to be dried out. Usually biocides are not suitable for sterilizing liquids, because there is no way of getting them out again afterwards.

- Irradiation. γ-Rays will sterilize anything, but are dangerous and relatively expensive to produce. UV light is an effective sterilizing agent, and somewhat safer: however to be sure that something is sterile it has to be exposed to UV for quite a long time (minutes to hours). In addition, UV does not penetrate very far into most liquids or solids, so it is usually only useful for sterilizing surfaces.

- Filtration. This is suitable for liquids or gases, but is extremely effective: usually a $0.2\,\mu m$ filter (i.e. a filter with holes in it about $0.2\,\mu m$ across) will remove all living things except viruses from a fluid.

Different sterilization methods must be chosen for different applications — there is no 'correct' method for all applications. The key problem to be overcome is materials compatibility. Thus many plastics are discoloured and rendered brittle by γ-rays, and melted by excess heat. Many fermentation and cell culture media cannot be autoclaved, as this would destroy some of the essential nutrient in them.

Strain (cultivar)

A strain of an organism is a type which is genetically distinct from other representatives of the species to which the organism belongs, but which is not different enough to be called a new species. Members of a strain are more genetically similar to each other than to members of other strains. The word 'strain' is normally used of microorganisms to describe a particular organism which has been isolated or engineered to have some property, like growing well or making a lot of a product. Isolating and improving strains of microorganisms is a major part of the process of making them suitable for an economic biotechnological process.

For animals, the term 'breed' or, sometimes, 'race' mean much the same thing — a genetically homogenous collection of animals, usually derived from one pair of parents, which is significantly distinct from other animals of the same species. Breeds or races can interbreed with each other, where animals of different species almost never can. Thus there are a large number of different 'breeds' of dogs (huskies, labradors, poodles, and so on) which can interbreed to form a 'generic' mongrel dog. For plants, the term 'cultivar' and 'variety' have a similar meaning. 'Strain' is sometimes used for plants, rarely for animals.

See also **Strain development, strain isolation**.

Strain development

Also strain improvement, this is the general term for improving the genetics of an organism so that it carries out a biotechnological process more effectively. The aims are to create a strain of the organism which makes what you want, makes it in large amounts, does not make much of anything else (so that you can purify your product easily), uses cheap and easily obtainable things to grow on, and does not require excessively careful control of culture conditions. The idea of an 'improved strain' is exemplified by coniferous trees used for wood pulp production: they grow almost anywhere on soil, air, and water, and you can make most of their mass into the product simply by pulping it. (This is why wood pulp is cheaper than, say, interferon.)

There are many routes to strain development.

- Incremental selection. This involves taking the current strain, treating it with mutation-causing chemicals (mutagens) and looking at a large number of the descendent strains to see if any have acquired a muta-tion which makes them more productive. This is a time- and labour-intensive operation, but frequently the most useful route to improving the production of chemicals such as antibiotics or amino acids from fermentations. This is a random screening approach, in which large numbers of random variants of an organism must be screened. Often the key to success lies in how these numbers can be screened rapidly and automatically, i.e. the system's 'screening capacity'.

The other methods are more directed.

- Hybridization. This is taking two strains and combining them genetically. It has been used extensively in agriculture, but, because the organisms used in biotechnology are so diverse, often cannot be used here so successfully. A variant which is more applicable to bacterial systems is the following.

- Conjugation. Here only a few 'desirable' genes are transferred between one strain and another.

- Genetic engineering. This seeks to alter the genetic make-up of an organism by directly introducing genes into it. These could code for a more efficient enzyme, or block the action of an enzyme which destroys the product you require. A more complex and costly path, but one which may be the only route open if 'traditional genetics' has failed.

Often the key to successful strain improvement via any route is finding a selection procedure. This is a set of conditions under which the strain you want has an advantage over all others. Finding a strain which makes an enzyme that breaks down one particular compound or group of compounds can be straightforward. For example, an oil-eating bacterium can be selected by growing a population of bacteria in a medium where the only carbon source is oil. Thus the only bacterium to flourish will be one that can metabolize the oil, and the faster it can metabolize it the faster it can grow. However, such relatively straightforward selection procedures are rarely available.

Strain isolation

This is the isolation of any bacterium, or indeed animal or plant, from the outside world. In general there are two approaches to strain isolation for microorganisms.

Large-scale sampling. Nearly all biotechnologically useful microorganisms are isolated from the soil, which holds between 1000 and 1 billion microorganisms per gram. The microorganisms which exist in a particular place depend on the local soil ecology, and clearly this varies greatly. Thus one approach to finding the ideal organism is to sample as

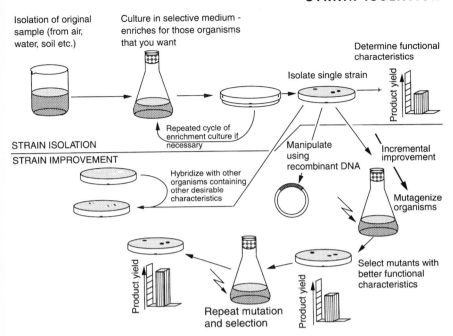

Isolation of original sample (from air, water, soil etc.)

Culture in selective medium - enriches for those organisms that you want

Determine functional characteristics

Isolate single strain

Product yield

Repeated cycle of enrichment culture if necessary

STRAIN ISOLATION

STRAIN IMPROVEMENT

Manipulate using recombinant DNA

Incremental improvement

Hybridize with other organisms containing other desirable characteristics

Mutagenize organisms

Product yield

Product yield

Select mutants with better functional characteristics

Repeat mutation and selection

Product yield

many soils as possible. Many major chemical and pharmaceutical companies have (or had) programmes under which any staff member travelling to exotic parts of the world brought home a small sample of soil to be used in the in-house screening programmes.

Location of appropriate environment. The other approach is to find an environment in which organisms will have had to develop the characteristic that is required in order to survive. Favoured sites are the effluent paths or waste tips for chemical plants, which tend to accumulate microorganisms that can break down all the chemicals that are present in their local environment. But many other possibilities exist. The organisms that break down methane, for example, were originally isolated from the soil around a cracked gas main.

Despite all the efforts that biotechnologists have gone to in developing recombinant DNA methods for optimizing bacteria for biotechnological uses, as often as not it is the original selection method that has the greatest impact on whether the organism is going to be the basis of a commercial process or not.

Strategic alliance

Not unique to biotechnology, this term means an alliance between two companies formalized as a legal agreement and usually aimed at developing some common area of business. Because setting up a complete biotechnology R&D department can be time-consuming and expensive, biotechnology companies and pharmaceutical companies frequently set up strategic alliances with each other to get access to know-how they would otherwise have to develop in-house. The partners might be after a stable source of funds and market support, a particular R&D technique, production facilities, product formulation and packaging ability, experience with regulatory authorities, or marketing and sales expertise. The value gained depends on which side of the alliance you fall: however the essence of an alliance is that both sides benefit while remaining independent.

Strategic alliances are different from specific research contracts (often called alliances, but actually normal contracts for one party to try to do something for another party — the only thing going from the contractor to the researcher is money), and mergers and acquisitions, where one partner usually loses autonomy. Probably the best-known Biotechnology acquisition/alliance of all was the acquisition of 60 per cent of Genentech by Hoffmann LaRoche in 1991. Genentech is probably big and energetic enough to retain its own identity, so making this a strategic partnership rather than the takeover that the balance-sheet makes it appear.

Substrate channelling

A neat idea which has been shown to work in research but not in large-scale application yet. The idea is to physically link together two enzymes which perform a series of reactions. The first takes substrate-1 and turns it into product-1. The second takes product-1 and turns it into product-2.

If both enzymes are added to a solution of substrate-1, then product-2 will accumulate. However quite a bit of product-1 will have to accumulate before there is enough for the second enzyme to work on. A faster and more efficient way of doing it is to link the two enzymes together physically, by making a fusion protein of them or by chemically linking them together. Then as soon as product-1 is made by the first enzyme, it is handed over to the second enzyme (which is right next door) and turned into product-2.

This has potential advantages in situations where product-1 is very unstable, or is liable to be acted on by other enzymes to turn it into an undesirable side-product. It is called substrate channelling because the process works as if there is a channel sending product-1 from enzyme to enzyme without it ever being released into the solution.

A related idea is to link a cofactor onto the enzyme. This has been done with the NADH cofactor of glucose dehydrogenase. As most dehydrogenases need NADH (or the related NADPH), if it is chemically linked to one enzyme then any other enzyme that wants to use that molecule has to come very close to the first to get its NADH. This, in effect, links the two enzymes together although they are not physically bonded most of the time.

Supercritical fluid enzymology

All materials have a critical temperature. (T_c) above which their gases cannot be turned into a liquid by compressing them. At this critical temperature, liquid and gas can coexist if the pressure is at the critical pressure (P_c). For carbon dioxide, for example, the critical temperature is 31°C. Thus at room temperature, if you compress carbon dioxide enough (as in a gas cylinder) it will turn into a liquid. Above 31°C, no matter how hard you compress it it will not liquefy — it will just become a denser and denser gas.

This highly compressed gas behaves in part like a gas, in part like a liquid. It is called a supercritical fluid (SCF), and has some useful properties for chemical and biotechnological processes.

- Diffusion in supercritical fluids is usually much faster than in liquids, so diffusion-limited reactions (which covers quite a number of enzymatic reactions) can occur faster.

- The solubility of chemicals in SCFs depends very sensitively on pressure. Thus reagents can be dissolved or products removed by precipitation, by altering the pressure. Some chemicals which are only sparingly soluble in water can be made extremely soluble in SCFs by choosing the right pressure and temperature.

- The pressures and temperatures involved are not damaging to many biopolymers.

SCFs have been used for several model enzyme reactions. In general it helps to include a small amount of water (which also dissolves in some SCFs) to assist enzyme stabilization: it is also essential if the enzyme uses water as a substrate.

Against the advantages, of course, is the disadvantage that SCFs must be kept at high pressure. One of the much advertised advantages of enzymes is that they work at mild temperatures and pressures. Working at 100 bar pressure in SCF removes one of these advantages. Thus SCFs are only useful for enzymatic catalysis if some other feature of using SCFs clearly compensates for the additional complexity of working with pressurized gas.

See also **Organic phase catalysis**.

Support

As a new technology with substantial potential economic impact, biotechnology has been supported by a range of governmental initiatives, especially in the US and in Japan. Some of the institutions interested in encouraging biotechnology are as follows.

OTA (Office of Technology Assessment). US central government agency which spots and advises on new technologies.

State Biotechnology Centers. 25 US states have centres which assist biotechnology. Usually set up on a university campus, they provide facilities for promoting links between academic and applied research and contacts with funding institutes, and promote their state's biotechnology in other states and other countries. They can also provide management expertise, and in some cases a venture capital fund or technical facilities.

In addition, many countries (and, in the US, many states) have incentives for new industry which benefit biotechnology. These include tax incentives (both local and national) and regulatory 'streamlining'.

See also **Clubs**.

Tank bioreactors

Bioreactors, also called fermenters, are the vessels in which fermentation takes place. Tank bioreactors are vessels in which the microorganism is grown in a large volume of liquid. This contrasts with fibre/membrane bioreactors and immobilized cell reactors. The large majority of bioreactors used in biotechnology are tank bioreactors, and most tank bioreactors are stirred tank bioreactors, because stirring helps to distribute gas and nutrient to the growing organisms effectively.

The bioreactor must provide a mechanism for introducing reagents and the microorganism into the reactor vessel, for providing substrate (i.e. food) to the microorganism (including oxygen for aerobic fermentation), for stirring it, and for keeping it at the right temperature, pH, etc. Temperature control is especially critical for large volume fermentations, as metabolizing microorganisms produce a great deal of heat. Variations in layout include different sizing and spacing of the baffles (which ensure that the volume is mixed thoroughly by the stirring) and different types of impellers (stirrers). These can be a wide range of shapes and sizes: disc (Rushton) turbine, open turbine, marine impeller (i.e. like a ship's screw).

The other main variation between reactors is the gas injection mechanism. This is almost always via a sparger (a pipe or plate with holes in it) which shoots bubbles into the base of the reactor), but a wide range of shapes of pipes or plates have been used. The shapes — which include rings, crosses ('spiders') or dead-end pipes — must be selected for the particular shape and size of the reactor, and the amount of gas that has to be injected.

There is a great deal of expertise in designing a suitable bioreactor for culturing an organism or cell type. As a consequence, there are more companies specializing in bioreactor design, control and engineering than in recombinant DNA techniques and reagents, despite the much higher profile of 'gene cloning'.

See also **Hollow Fibre, Immobilized cell bioreactors.**

Targeted drug delivery

This is using any method to deliver a drug to the site in the body where it is needed, rather than allowing it to diffuse into many sites. There are three approaches to such targeted drug delivery.

The first is to encapsulate the drug in something, usually a lipid coat (i.e. a liposome — *see* **Liposome**). The outside of the coat is itself coated with a material which binds to the target cells — an antibody specific for those cells, a glycoprotein, or a receptor molecule or ligand. The liposome travels around the blood until it finds its target: once there it sticks to the cell, the lipid membrane fuses with that of the cell, and releases the contents into the cell.

The second approach links the targeting mechanism directly to the drug. Here the drug must either operate outside the cell, or be able to get itself into the cell. A much-talked about application is linking toxin proteins to antibodies: the proteins can get inside cells and there destroy the cellular machinery, but only when carried near enough to the cell by the antibody. Such conjugates are called immunotoxins. The application here, clearly, is to destroy cancer cells or, conceivably, cells infected with long-term viruses such as HBV.

The problem with both these approaches is how to get the drug–carrier complex from the blood stream into the target tissue: unless the target is the endothelial cells of the blood vessels or a few cell types in liver, lung, or kidney, nothing as large as a liposome is going to be able to escape from the blood vessels to get to them.

The third approach is to make the drug as a 'prodrug' which goes to all tissues of the body, but which is metabolized to the active drug only by one tissue because that tissue has a high level of an enzyme which cuts the prodrug into inert 'carrier' and active drug. This is easier to do for tissues such as liver and kidney which have a battery of rather specific enzymes.

See also **Immunoconjugate, Immunotoxins**.

Thermal sensors

Thermal sensors, sensors that detect tiny changes in heat or temperature, are well known in many applications. Such sensors are often used in gas chromatography systems to detect molecules from the GC column. There have been some attempts to harness thermal sensors as **biosensors**. Here a probe detects the heat given out when an enzymatic reaction occurs. This could be much more flexible than **enzyme electrodes**, as, while relatively few enzyme reactions involve the transfer of electrons which could be picked up by an electrode, nearly all result in the release of heat. The problem is that, for small samples of dilute material, the amount of heat released is tiny, so very small and very sensitive heat sensors are needed.

Thermophile

A thermophile is an organism which grows at a higher temperature than most other organisms. Generally, as a wide range of bacteria, fungi, and simple plants and animals can grow at temperatures up to 50°C, 'thermophiles' are considered to be organisms which can grow above 50°C. They can be classified fairly arbitrarily depending on their optimal growth temperature into 'slight thermophiles' (50–65°C), 'thermophiles', (65–85°C) and 'extreme thermophiles' (>85°C). Thermophiles and extreme thermophiles are usually found growing in very hot places: hot springs and geysers, smoker vents on the sea floor, and domestic hot water pipes for example.

Thermophiles are of interest to biotechnologists because of the economics of fermentation and biotransformation. Many industrial processes could be catalysed by enzymes, but the enzymes are too slow. They may be speeded up by heating the reaction up, but this rapidly destroys the enzyme. Heating up the reaction is also desirable because it reduces viscosity and increases the rate of diffusion of reagents, and so reduces the amount of stirring and pumping energy needed, and the heat prevents other enzymes from working or (usually) contaminating organisms from growing in your reactor.

Enzymes from thermophiles can, of necessity, withstand such high temperatures. They also frequently exhibit increased stability to organic solvents. Thus there is substantial interest in isolating these enzymes and using them in industrial processes. Because the bacteria themselves are usually tricky to grow (and must be grown at high temperatures), once a suitable enzyme is identified it is common to seek to 'clone' its gene into a bacterium which grows at a more moderate temperature. This also means that they may be purified from all the other proteins in the bacterial cell simply by heating it up: all the other, non-thermostable proteins will precipitate, leaving a reasonably pure preparation of the target enzyme.

A range of thermostable enzymes are used in industrial processes. As with all research into isolating enzymes from bacteria, one key feature is to have a large number of diverse sources of candidate organisms to screen. This is why Iceland, with one of the world's densest concentrations of different types of hot spring and geyser, has been the source of a majority of the publicly available thermophilic organisms in use.

Tissue culture

This is sometimes used interchangeably with 'cell culture'. Strictly it means the cultivation of tissues, i.e. multicell assemblies, outside the body. However it is often used to describe cell culture — the culture of isolated cells outside the body — as the two use very similar techniques and materials.

The requirements for cell culture are simple to state, harder to get to work. The principle condition is sterility — yeasts and bacteria grow much faster than cultured cells, and so if just one bacterium gets into a cell culture it will soon outnumber the mammalian cells. The bacterium's metabolic wastes, and particularly the acid it produces, will then kill the cells. Thus other organisms must be excluded rigorously. This is relatively simple to do on a laboratory scale, but harder if the cells are to be produced in bulk.

Other conditions relate to the medium in which the cells must live. This must contain a wide variety of nutrients including vitamins and amino acids, and growth factors to stimulate the cells to divide. In the laboratory these are provided by serum, usually fetal calf serum (FCS), but this is much too expensive to use on a production scale and so a variety of nutrient additives, lipids and lipoproteins, and peptide growth hormones have been made to encourage mammalian cells to grow. The peptides needed vary between different cell types (which is why FCS is so commonly used in research — it has almost every growth factor in it).

A key parameter in cell culture is whether the cells are anchorage dependent or anchorage independent. The former means that the cells have to stick down to a surface to grow: the latter that they can drift free in solution. Sometimes anchorage independent cells will stick onto things anyway, but they do not need to in order to survive.

Mammalian cell culture is used widely in biotechnology. Monoclonal antibodies are manufactured in bulk using cell culture (*see* **Monoclonal antibody production**). A range of biopharmaceutical products are produced in genetically engineered mammalian cells, as these synthesize the correct glycoforms of the proteins.

Tissue culture differs from cell culture in that tissues isolated from animals are mortal, as are cells isolated directly from animals. By contrast, cell lines are immortal in the sense that they will go on growing and dividing indefinitely (*see* **immortalization**).

Toxins

Living things make some of the most dangerous compounds known that are not radioactive, like ricin (castor bean toxin) and pertussis (whooping cough) toxin. One molecule of botulinus toxin, delivered to the inside of a cell a billion times more massive than itself, will kill the cell. Such powerful poisons have potential uses, and biotechnologists have the potential to make them relatively safely.

Toxins can be used on their own as therapeutics. Botulinus toxin is being developed as a way of blocking unwanted muscle spasm. Clearly it cannot be injected generally as other drugs would be — it would kill the patient. However if minuscule doses are injected into the muscle they can paralyse that muscle. The amount of protein used is so small that it escapes the notice of the immune system, and so the body does not make antibodies which could neutralize future doses. Allergan and Porton International are producing a commercial version of botulinus toxin for this application.

Toxins can also be coupled to other things to give them a lethal 'sting'. Immunoconjugates are probably the best example (*see* **Immunoconjugate**).

Making such toxins is difficult, even with all the variety of microbiological containment methods available. People have tried cloning the genes for these protein toxins into bacteria to express them more efficiently (as they are normally present in extremely small amounts). Such scientists tend to find themselves in the middle of empty rooms whenever they talk about their ambitions at conferences.

Transfection, transduction, transformation

These terms are all used to mean ways of getting DNA into cells, usually animal or bacterial cells. The meanings are different depending on the type of cell being discussed.

- Transfection. Strictly, this means carrying a piece of DNA into a cell as part of a virus particle. For mammalian and plant cells, it is used more generally to mean almost any way of getting DNA into a cell.

- Transduction. Not used much any more, this means transferring a piece of DNA from one organism to another via natural DNA exchange processes. It occurs almost exclusive in bacteria, and is a method for genetically engineering large pieces of DNA such as the *Agrobacterium tumefaciens* Ti plasmid.

- Transformation. For bacteria, this means getting the bacterium to take up DNA which the experimeter has added to its medium. The bacteria able to do this are called 'competent'. Demonstrating transformation was one of the key proofs that DNA was the genetic material. For plants, transformation is used to imply the stable integration of a foreign DNA into the plant genome. This is often via the *A. tumefaciens*-based transformation system. For mammalian cells, transformation means turning the cell from a cell whose growth is limited by its neighbouring cells into one whose growth is limited only by the media available to it. Transformation is a step in the development of cancer cells, and is also a crucial step in generating an 'immortalized' cell line. Because these two meanings of 'transformation' grew up beside each other, genetic engineers who are manipulating mammalian cells often say that they 'transfect' the cells with DNA, rather than 'transform' them, even if what they actually do is add naked DNA to the cells.

There are several common methods used to place 'naked' DNA (i.e. DNA that is not encapsulated in a virus particle, a liposome, or some other carrier system) into cells.

- Bacterial cells. Bacterial cells that are 'competent' (i.e. in a suitable physiological state, which may be achieved by growing them in the right way and suspending them in the right buffer) will spontaneously take up DNA from the solution around them. The common factor involved is usually the need for magnesium salts in their medium.

- Bacterial protoplasts can also be transformed by fusing them together in the presence of DNA. This can be done using polyethylene glycol (PEG). The cells' membranes join up in the presence of PEG forming poly-cell masses, and some of the external solution, containing the DNA, gets trapped inside the cell in the process.

- Mammalian cells can be transfected by adding DNA to them as a calcium phosphate precipitate.

See also **Biolistics, Electroporation, Retrovirus**.

Transgenic

A transgenic organism is one which has been altered to contain a gene from another organism, usually from another species. While this would suggest that any genetically engineered organism could be called 'transgenic', the term is usually only applied to animals. Bacteria and yeasts are always called 'genetically engineered', and with plants it is an even choice which term to use.

Creating transgenic plants is a relatively mature science (*see* **Plant genetic engineering**).

Creating transgenic animals is more complex. The germ cells (i.e. the egg, sperm, or newly fertilized zygote) must be altered — altering some of the cells in the adult (the somatic cells) is no help at all (although it may be valuable for other reasons.) Thus, unlike plant genetic engineers who can regenerate a new plant from almost any cell in the plant, animal genetic engineers must develop methods for getting DNA into the germ cells. There are several ways of doing this:

- Microinjection. This, the first successful method, simply injects the DNA into the egg's nucleus (diameter about 1/100 mm) with a very fine needle. Microinjection needs considerable skill. This is the only method which works for cows, sheep, goats, and pigs.

- Transfection. This is chemical treatment of the egg with the DNA. While this works well for somatic cells, it is 'a bit dodgy' for eggs. An Italian group claimed that they had found a simple way of making sperm absorb DNA from solution — however no one else has been able to repeat their results.

- Electroporation (*see* **Electroporation**). Not very successful with animals cells, and not at all successful with eggs.

- Using embryonic carcinoma cells (EC cells) to create a chimera.

- Retroviral vectors. Some viruses, notably the retroviruses, can carry DNA into a cell and splice it into the cell's own DNA. There is a lot of interest in harnessing this ability to genetically engineer all sorts of animal cells.

- Transomics. This is an injection technique, but instead of injecting pure DNA its practitioners dissect out sections of chromosome under the microscope and inject them. As chromosomes are only a 1/1000 mm or so long, (and much thinner), this is not a profession for myopics or those with shaky hands.

Foreign genes introduced into transgenics are usually called exogenous (in animals) or ectopic (in plants) genes.

See also **Chimera, Gene therapy, Transgenic animals: applications.**

Transgenic animals: applications

There are three areas in which transgenic animal technology has been used to create biotechnological products, as opposed to research results.

The first is creating animal models for disease. This is probably the most successful application to date (*see* **Transgenic disease models**).

The second is to use animals as expression systems for protein manufacture, especially for producing biopharmaceuticals. The aim is to genetically engineer animals so that they contain the gene for the biopharmaceutical spliced onto a promoter and a signal peptide which makes them express the protein in the mammary gland — the engineered protein is then made in the milk. Protein levels of up to $3 \, \text{gl}^{-1}$ have been reported. Pigs, cows, sheep, goats, and rabbits all have their enthusiasts for this technology. The advantages over fermentation production systems are that: the need for sterile culture is avoided; the need for complex nutrient mixes is avoided; and the protein is obtained relatively free of other proteins, and quite free of cell wall materials or potential endotoxins. This technology has been called 'pharming', although only by journalists.

Several groups of researchers have made transgenic animals that produce milk containing several grams per litre of alpha-1-antitrypsin, a protein with potential for treating emphysema. Pharmaceutical Proteins Ltd. have used sheep, and Genzyme and Tufts University have used goats to make this protein. The original idea of using cows (traditionally associated with making milk) has lost favour because their longer breeding cycles and small numbers of offspring made breeding more costly and time-consuming.

The third application area is in farm animal improvement. Sixty per cent of the fixed cost of producing a pig is the cost of feed, so if a pig could be engineered to turn that feed into meat more efficiently, that would represent a substantial saving for the farmer. In principle, expression of a transgenic growth hormone gene in the pig should do this: however in experiments to date the side-effects of engineering growth hormone genes into pigs or cattle have outweighed the potential benefits. In addition, the debate over the use of injected **BST** suggests that, even if the genetic engineering is successful, it will be controversial on regulatory and social grounds.

Other ideas for engineering of farm animals have included improving wool quality, and milk quality by introducing more milk protein into cows' milk.

See also **Wool, Yuk factor**.

Transgenic disease models

One application of transgenic animals is to model human diseases. When human patients for a disease are rare, when it is impossible to find them before the disease is well established so that the early stages cannot be studied, or when it is not ethical or practical to do studies on humans, having an animal model of the disease is essential. However quite a few human diseases are not mimicked accurately by any animal model.

Transgenic technologists seek to create animals, especially mice, that get diseases which are in some specific way characteristic of a human disease. These can then be used for screening for potential new therapies or drugs. Among the models used are the following.

Humanized ('humanoid') mice for AIDS research. True transgenic mice with the human CD4 gene can be infected with the AIDS virus. Another model — the Hu-SCID mouse — has no functional immune system of its own, but has human immune cells inserted into it to make an immune system which is susceptible to AIDS. (This is properly called a chimeric animal, because it is a mixture of cells or tissues from several different animals.) SCID mice can be made in several ways that knock out their immune system, including massive whole-body irradiation and genetically engineering them to include a toxin gene which is expressed at high levels in their lymphocytes.

Models for diabetes (and several other diseases in which specific cells are absent or do not work properly). A toxin gene spliced onto a promoter sequence which only expresses that toxin gene in one particular tissue is placed in the animals. In the case of diabetes, the toxin is expressed in the beta-cells of the pancreas. The toxin then kills those cells off, leaving the rest of the animal healthy. Such genetic constructs are called toxigenics.

Models for cancer. Models for cancer usually have an oncogene inserted into them, so that they develop a specific cancer at an anomalously high rate.

Models of immune function. A key aspect of the healthy immune system is its ability to distinguish the normal constituents of the body from other, potentially invading materials. A wide range of diseases arise from a failure of this mechanism. Transgenics are being used to find out how the immune system 'learns' to discriminate self from non-self, both by introducing foreign protein genes into mice and by creating toxigenics which lack certain sets of lymphocytes. These studies have implications for many diseases, such as diabetes (which has an autoimmune component), arthritis, allergies, and multiple sclerosis.

Another approach is to use homologous recombination to disrupt a gene in the animal, and so model directly a human disease in which a gene is faulty. Several of the collagen diseases such as osteogenesis imperfecta have been modelled in this way.

See also **Homologous recombination, Oncomouse**.

Transmissible encephalopathies

This is a generic term for the diseases bovine spongiform encephalopathy (also 'mad cow disease') scrapie, Krutzfelt–Jacob syndrome, kuru, transmissible mink encephalopathy. They are slow, degenerative diseases of the brain. No one knows what causes them, although it seems likely that a protein called the prion protein is responsible. The causative agent is very hard to destroy: boiling it, digesting it in acid, or leaving it in the sun for a week seem to have little effect.

The encephalopathies started to cause concern in the biotechnology industry because of the possibility that the agent that causes the disease, whatever it is, will get into biotechnology products produced from cultured cells. Many cell culture systems use fetal calf serum as part of the medium in which the cells grow. The fear is that the scrapie/BSE agent could get into the cells, and from there into the biotechnology products. The Dutch Health board refused approval of the Ares–Serono growth hormone on this basis in 1990.

Transposon

A transposon is a genetic element that can move around the genome. Most genes stay where they are with respect to other genes, unless mutation rearranges the genome in which they reside. Transposons break this rule. They are able to create copies of themselves elsewhere in the genome, or even in other genomes if they are present in the same cell. Thus, for example, a transposon might copy itself out of a bacterial genome and into the genome of a bacteriophage as the phage infects the bacterium. Some transposons splice themselves out of their original site to do this, but most simply copy themselves, so the end result of the copying process is two copies of the transposon where one existed before.

The process of a transposon moving is called transposition. It has been harnessed in a variety of ways by geneticists and genetic engineers to move genes around in bacteria and to a lesser degree in plants.

Many transposons carry useful genes as well as being 'selfish' DNA that propagates itself around the genome. Most antibiotic resistance genes are carried on transposons in some bacteria, as are genes for things like heavy metal resistance.

The way that many transposons move is reminiscent of the way that retroviruses reproduce, in that the transposon is transcribed onto an RNA which is then copied back into the genome as a DNA. Because of this similarity, such transposons and retroviruses are sometimes grouped together and called retrotransposons.

Treatment protocol program

This is an FDA initiative to allow terminally ill patients to be given experimental drugs before they have cleared all the hurdles of full regulatory approval. This concept was driven to public notice by AIDS patients, who objected that the rate of approval of new drugs for AIDS was so slow that they would be dead before anyone got a drug on the market to cure their disease.

See also **Drug development pathway, Regulatory authorities (US)**.

Triple DNA

Most introductory textbooks will tell you that RNA is 'single stranded' and that DNA is 'double'stranded', that is, that DNA consists of a helix of two strands wrapped round each other. However it has been known for a while that RNA can be triple stranded, and recently triple stranded DNA has been demonstrated too (also sometimes called triplex DNA). It has several potential applications.

The 'third strand' of triple DNA binds to the other two by specific base pairing, and so it can be used as a reagent which recognizes a specific DNA sequence. If it is linked to a molecule which cuts DNA, the third strand can be shown to act as a sequence-specific nuclease, i.e. a reagent which will cut the DNA at (or just next to) a very specific site. Several 'artificial nucleases' of this sort have been made.

Alternative uses include using it to block gene activity in much the same way as **antisense** RNA does, by binding to genes and so blocking their transcription. Aptamers are DNA molecules selected for their ability to bind to genes and so potentially to block their activity.

A third area of possible interest is as a DNA probe in disease testing — using a third DNA strand to form a triple helix means that you do not have to unwind the other two before doing a hybridization test.

There are a number of related complex structures which have been made from DNA, for various purposes. Chiron made branched polymers of DNA as an aid to increasing the sensitivity of hybridization assays. Nadrian Seeman has used oligonucleotides to make cage-like structures, opening the intriguing possibility of using DNA as a **biomaterial**.

See also **Darwinian cloning**.

Tumour marker

A tumour marker is any molecule which shows that a cancer is present. Usually they are produced by only a few types of cancer, so as well as showing that a cancer is present they help to identify the type of cancer, and hence the most suitable therapy.

Tumour markers are of interest in biomedicine because of the importance of cancer as a cause of death in the Western world. Tumour markers can be used for diagnosis or, potentially, as the targets for biopharmaceutical drugs such as **immunotoxins**.

Tumour markers fall into two categories. The first type are the products of oncogenes, and hence their presence represents part of the reason why the cell is a cancer cell to start with. The others are incidental, but are always found associated with a particular type of cancer. Such proteins are usually made in only a few cell types in the healthy body, but cancer cells make them in much larger amounts or in inappropriate places. Among the more talked about are:

- Beta-2-microglobulin

- CEA (carcino-embryonic antigen: a protein found on many cancer cells and in normal embryos)

- NSE (neurone-specific enolase, an enzyme usually only found in nerve cells)

- AFP (alpha-fetoprotein, a protein normally only made by the developing embryo)

- HCG (human chorionic gonadotrophin, a protein usually made only by the placenta)

- EMA (Epithelial membrane antigen)

- CA 125, CA 19-9 (two cell surface proteins found on several cancers of the female reproductive tract: no one knows what they normally do)

- TPA (tissue polypeptide antigen. Nothing to do with tissue plasminogen activator, the heart drug!)

- PAP (prostate acid phosphatase, an enzyme which is a 'marker' for prostate cancer.

In addition there are a wide range of antigens (i.e. proteins to which antibodies bind) which have been identified by monoclonal antibodies as being associated with specific types of cancer, but whose normal functions are obscure. A number of them are glycoproteins or carbohydrates: cancer cells often add sugar units to proteins in a slightly different order from normal cells, so creating different glycoforms of those proteins: it is these differences between the glycoforms which are detected as 'markers' by the antibody.

See also **Glycosylation, Oncogene**.

Vaccines

Vaccines are preparations which, when given to a patient, elicit an immune response which subsequently protects the patient from infection by the disease-causing agent. Usually the vaccine consists of the organism which causes the disease (suitably attenuated or killed) or some part of it. Attenuation of a virus (or bacterium) is growing it so that it does not lose its ability to grow in culture, but does lose some or all of its ability to cause disease in animals. Usually bacteria, and to a degree viruses, slowly lose their ability to colonize living things (and hence to cause disease) as they are cultured outside the body. There are a range of biotechnological approaches to producing vaccines.

- Viral vaccines. These are vaccines which consist of genetically altered viruses.

- Biopharmaceutical vaccines. Proteins, or sections of proteins, which are the same as the proteins in a virus's or a bacterium's wall, can be made by **recombinant DNA** methods as vaccines. This is a standard biotechnological route, and has the advantage that there is no possibility that the resulting vaccine will contain any live virus particles. Peptide vaccines are often incorporated by genetic engineering into a larger carrier protein to improve their immunogenicity (i.e. how well they cause the body to become immune), or their stability.

- Multiple antigen peptides (MAPs). Developed by J. T. Tam, these are peptide vaccines which are chemically stitched together (usually onto a polylysine 'backbone'). This means that several vaccines can be delivered in one shot.

- Polyprotein vaccines. This is a similar idea to MAPs, but here a single protein is made, by genetic engineering, in which the different peptides form part of a continuous polypeptide chain.

See also **Viral Vaccines**.

Vaccinia virus

Vaccinia viruses are DNA viruses from the same family as cowpox and smallpox. However they are safe viruses to work with, and so have been used for several biotechnology applications.

Specific vaccinia viruses have been used as the basis of an expression vector system (*see* **Expression systems**). The virus can infect a wide range of cells, and has a lot of DNA, quite a bit of which can be removed using suitable genetics. Thus quite large foreign genes can be spliced into it, and then the recombinant virus can be used to infect a wide range of cells, allowing the biotechnologists to choose the cell most suitable for the process. Vaccinia virus vectors have been used quite widely in research, as they can be used to express proteins in mammalian cells. Because they can hold quite a lot of DNA, they can be used to make more than one protein at once in a cell, which can be useful for making proteins with more than one polypeptide chain (multi-subunit proteins).

Vaccinia has also been used as the basis for 'live virus' vaccines (*see* **Viral vaccines**). It is suitable for this because it does not itself cause severe disease, and because it can infect a wide range of species it can be used to produce a range of animal vaccines, which are the first targets of this sort of technology. Provisional approval for a field trial of a live vaccinia virus vaccine was granted in the US in 1990.

Foreign genes are usually introduced into vaccinia by recombination, rather than by isolating the vaccinia DNA and manipulating it *in vitro*. This is because the vaccinia virus is too large to be manipulated conveniently.

Cowpox and racoonpox viruses, which share some of the useful characteristics of vaccinia, are being looked at as alternative vector systems.

Vector

A vector in biotechnology is usually a DNA segment which allows another piece of DNA to be 'cloned' using **recombinant DNA** techniques.

DNA does not replicate all by itself: it needs a battery of enzymes to replicate inside a cell. The enzymes co-ordinate DNA synthesis with cell growth by only starting the synthesis of a DNA molecule at a specific time in the cell's growth cycle. To allow this to happen the DNA must contain a 'start here' signal, called the origin of replication. Thus any DNA which is to be cloned must contain an origin. A unit of DNA that has an origin of replication (and a signal to stop replication at the other end, if that is necessary) is called a replicon. As most bits of DNA do *not* contain an origin, they must be given one: this is done by splicing them together with an origin-containing bit of DNA, called a vector. Vectors can be thought of as little replicons on to which we can add other DNA. This is the basic function of vectors. To make them convenient to use, they have a range of other properties.

Most cloning vectors are episomes, that is they are genetic elements which can be replicated separately from the host cell's chromosome (i.e. the rest of its DNA). Episomes can be plasmids (small loops of DNA with no function that is deleterious to the cell) or persistent viruses (bits of DNA with the potential for coding for virus particles) (*see* **Plasmid**). 'Traditional' vectors such as the pBR series and the 2-micron vectors for yeast are plasmids, which the lambda-series of DNA sequencing vectors is based on a bacteriophage (a bacterium-eating virus). Other viruses such as T7 are also used, and bits of them have been used in the construction of more exotic beasts such as cosmids: these, much used in large-scale gene cloning, are plasmids which can be packaged up into lambda-virus particles, but only when you have put 40 000 bases of foreign DNA into them. Thus the packaging process is an excellent way of ensuring that you have a plasmid with a large stretch of DNA inserted into it.

Vectors contain a range of genetic elements to make cloning with them easier. These can include:

- Selectable genes. These code for something which allows the cell to survive an otherwise hostile environment. A common one is a gene for resistance to an antibiotic: growing the engineered organism in the presence of antibiotic will select those organisms which contain the vector (and hence whatever genes we have spliced into the vector).

- Polylinker. This is a piece of DNA made to contain many restriction enzyme sites, so that the vector can be cut at that one place for splicing into other genes.

- Other origins of replication. Origins are specific to the type of organism — bacterial ones usually do not work in yeast. Specific organisms are useful for different parts of a genetic engineering project, so some vectors contain origins of replication for several organisms. Such vectors are called shuttle vectors because they can shuttle between species (with the scientist's help).

- Specialist origins. Other variations on origins of replication are

  — High copy-number plasmids, which exist as many copies inside a cell, not the usual one or two.

  — Runaway replication plasmids, where, on some trigger signal (usually a change in temperature) the normal controls on how much plasmid DNA there is in a cell break down and the cell fills up with the plasmid.

- Promoters, enhancers, leader peptides. These are elements to help you express a gene that is cloned in the vector.

Because there are so many possible vectors which can be assembled out of these components, some vector systems are made not as complete vectors but as cassette systems, where different selectable genes, origins, etc. can be slotted together to make the vector of your choice.

See also **Expression systems**.

Vertical integration

A 'must' term for management consultants, this means a company which can perform all parts of development, production, and sale of something.

In the pharmaceutical industry, a vertically integrated company would be one which researched, developed, manufactured, marketed, and sold a drug.

There is a substantial difference between the vertical integration levels of US and European biotechnology companies. Many US biotechnology companies with pharmaceutical links usually see themselves as providing a service for larger, 'mainline' pharmaceutical companies: they discover or invent drugs, develop new ways of delivering them, or provide research or development capabilities for making the drugs. By contrast, most European biotechnology companies feel that it is their destiny to *become* large drug companies, doing everything from drug discovery to knocking on family doctor's doors. (This may be one reason why there are many fewer biotechnology companies in Europe than in the US.)

In other parts of the healthcare industry, biotechnology companies tend to stay away from trying to become another Glaxo or Dow. Outside healthcare, in areas such as environmental cleanup or specialist chemicals synthesis, the same criteria do not apply, as the biotechnology companies act as service suppliers for other companies and individuals in many industries. Some, especially those which provide chemicals to the drug industry, also have the lure of being fully integrated drug companies — again, there is a tendency for European companies to take the big, long-term view (or have delusions of grandeur, depending on your ambitions), while US companies carve out a niche role serving existing drug companies.

Viral vaccines

Also called live virus vaccines, these are vaccines which consist of live viruses rather than dead ones or separated parts of viruses. Clearly, the virus itself cannot be used, as that would simply give the patient the disease, so instead, one of two genetic engineering methods is used, to produce a virus which will elicit an immune response to a viral pathogen but which will not cause the disease itself.

The first method is to genetically engineer the disease virus so that it is harmless, but can still replicate (albeit inefficiently, sometimes) in cultured animal cells. This is similar to producing an 'attenuated' virus, i.e. one which has been grown in the laboratory until it loses its ability to cause disease. However, the genetic engineering route seeks to make sure that the attenuated virus has no chance of mutating back to a 'wild-type', pathogenic virus, by deleting whole genes or replacing key regions of genes with completely different genetic material.

The second approach is to clone the gene for a protein from the pathogenic virus into another, harmless virus, so that the result 'looks' like the pathogenic virus but causes on disease. **Vaccinia** and **adenoviruses** have been used in this way, notably to make rabies viruses for distribution with meat bait: a trial of such a vaccine was carried out in summer 1990 in the US.

Walking

There are several techniques known as 'gene walking' or 'chromosome walking'. They are all methods for cloning large regions of a chromosome. The diagram illustrates the basic idea. Starting from a known site, a **gene library** is screened for clones that hydridize to DNA probes taken from the ends of the first clone. These clones are then isolated, and *their* ends used to screen the library again. These clones are then isolated and *their* ends used... and so on. This can go on as long as is needed to get from where you are (usually at a 'linked marker' — an **RFLP** site known to be near the gene of interest) to where you want to be.

There are variants called 'gene jumping' or 'chromosome jumping' that allow some of the intermediate steps to be cut out: these rely on rearranging the original chromosomal DNA during cloning.

To 'walk' a chromosome quickly, it is useful for the clones to each cover an extremely large amount of DNA, as otherwise each 'step' will only cover a small amount of the genome. Thus cosmid vectors (which hold 40 000 bases of foreign DNA per clone) and YAC vectors (which can hold up to a million bases) are preferred (*see* **Vector, YACs**).

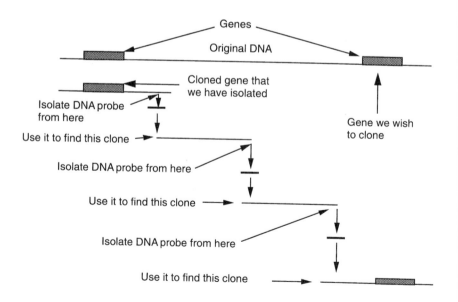

Wood

Wood processing is attracting increasing attention from biotechnologists, in part because traditional methods produce a lot of effluent that is considered to be environmentally unfriendly and in part because wood is a biological material that is well suited to processing by biological means. Nearly all the bioprocessing of wood is aimed at paper production, which takes wood chips and turns it, via wood pulp, into clean, white cellulose for making paper.

The five areas on which biotechnologists have focused are:

- Pre-processing. This is the removal of pitch and resins from wood. The wood from most trees contains a substantial number of complex, oily chemicals which preserve the wood from being attacked by insects and bacteria. These have to be removed: this can be achieved by 'fermenting' the wood pulp with microorganisms which grow on pitch or by digesting it with lipases that break the pitch down into water soluble materials.

- Pulping. Usually wood chips are converted to pulp mechanically or by using chemicals. Enzyme methods are being investigated. The object here is to break down the lignin and other non-cellulose materials that hold the cellulose fibres together. Several fungi that make ligninases are known, and these can contribute to breaking the wood down. At the moment such methods are used in conjunction with mechanical mashing. Enzyme or fungal treatment softens the wood, reducing the power needed by the mechanical pulpers.

- Fibre modification. The nature of paper depends strongly on the sort of fibres that make it up. The cellulose fibres can be modified by trimming off surface irregularities.

- Biobleaching. The colour of paper is extremely important. Wood is coloured because of a large number of compounds that permeate the fibres, primarily materials that fall under the general heading of 'lignin'. Ligninases have been used to bleach pulp without resort to the chlorine and chlorine oxides usually used in the paper industry. Xylanases are also used here: these break down polysaccharides other than cellulose and so release coloured materials trapped in the pulp. (It is important that the xylanases be free of any contaminating cellulose, as this would break down the cellulose as well.)

- Waste disposal. Producing new paper and recycling old paper generates a lot of waste water that contains a lot of carbon compounds. These can be a substantial pollution problem, raising the BOD (biological oxygen demand) of the waste water to unacceptable levels. Biological treatment of wood pulp waste is a way of reducing these effluent problems.

Wool

One of the targets of the application of genetic engineering to farm animals is to improve the quality or quantity of wool that a sheep produces. This is a complex problem, but one that a number of research groups are working on, especially in Australia, which produces a substantial fraction of the 2 billion kg of wool produced annually world-wide.

Improved wool production focuses on the following.

Inserting the gene for growth hormone in sheep. This has been tried, and seems to produce an increase in wool production, although no one is sure why.

Inserting new genes for keratins into sheep. There are several types of keratin in wool, and altering their ratio may improve the quality of the wool. This is an empirical approach, because it is not clear what effect any particular gene insertion will have on the wool even if it makes protein in the right cells and at the right time.

Inserting the genes for improved synthesis of cysteine into transgenic sheep. Keratin, the protein in wool, has a lot of cysteine, which is a limiting factor in the rate of growth of wool. Sheep cannot usually make cysteine themselves since they lack the relevant enzymes, so the engineering aims to give the sheep enzymes from bacteria, which can make cysteine from sulphides generated in the rumen.

Engineering feed plants. An alternative method to getting more cysteine into sheep is to engineer the plants they eat to have more cysteine. The problem here is that the rumen bacteria break down a lot of the cysteine in the food, and so improving the forage plants may not improve wool output. Some storage proteins from peas are proof against rumen breakdown, and may be suitable.

Engineering rumen bacteria. An alternative route is to manipulate the rumen bacteria to convert cellulose in the feed into chemicals that the sheep can use more efficiently, or to make more essential amino acids, especially cysteine, available to the sheep. This research is still in a very early stage, in part because, to model accurately what the bacteria do, you need something very like a sheep's stomach as an incubator.

Xenobiotics

A xenobiotic is a chemical which would not normally be found in a given environment, and usually means a toxic chemical which is entirely artificial such as a chlorinated aromatic compound or an organomercury compound. Biotechnology brushes with xenobiotics in three areas. Firstly in determining their toxicity and effects on living systems. Secondly, biotechnologists have developed methods for removing them through **bioremediation** or enzyme-based degradation. Lastly, a range of biotechnological products aim to replace the compounds which, if they get out of their target site, are classified as xenobiotics. Among these are the chemical herbicides and pesticides which biocontrol agents and biopesticides hope to replace.

YACs

YACs are yeast artificial chromosomes, and are cloning vectors which are gaining a lot of use in the human genome project (*see* **Genome project**).

They consist of those pieces of DNA which define the ends (telomeres) and the middle (centromere) of a chromosome in yeast. Both of these elements are needed to allow a chromosome to be replicated in yeast cells: if there is no telomere then the ends of the chromosome are liable to be broken off, or to join onto other chromosomes. If there is no centromere, then newly made chromosomes will not be pulled into the new cells during cell division. In addition, there is an origin of replication so that the DNA is replicated.

These elements are placed in a single DNA fragment, which can be used as a vector to clone foreign DNA into yeast. The advantage of a YAC is that there is no effective limit to how big the piece of DNA can be. Thus, while conventional bacterial cloning using bacteriophage or plasmids is usually limited to cloning foreign DNA fragments of a few tens of thousands of bases long, YACs can and have cloned fragments millions of bases long. This makes mapping whole genomes of DNA much easier, as the whole genome map has to be assembled from far fewer YAC maps, and also makes cloning very large genes such as the gene for muscular dystrophy (which is at least 2 million bases long) more straightforward.

However there is nothing that can be done with YACs which cannot be done with some ingenuity using other vectors (*see* **Yeast cloning vectors**).

Yeast cloning vectors

After a few bacteria, the yeasts, and especially *Saccharomyces cerevisiae*, are favourite organisms in which to clone and express DNA. They are eukaryotes, and so can splice out introns, the noncoding sequences in the middle of many eukaryotic genes. They also carry out some **glycosylation**, although not usually in the same way as mammalian cells. Also, because they are not bacteria, they produce far fewer endotoxins which have to be removed from recombinant protein products. They also grow very quickly compared to mammalian or insect cells, which enables large amounts of them to be prepared more easily and reduces problems with contamination, as far fewer organisms can outgrow them.

Among the vectors used to clone DNAs in yeast cells are:

- YACs (yeast artificial chromosomes). Very popular with the **Genome project**, as they can 'clone' very big bits of DNA.

- 2-Micron plasmids. The 2-micron circle is a naturally occuring yeast plasmid. It has been used to form the basis of several cloning vector systems. Also called YEps (yeast episomal plasmids).

- YIp (yeast integrating plasmid). A plasmid which inserts itself into the DNA of one of the yeast's chromosomes. Genes integrated into the yeast chromosomes are less liable to be lost by the yeast as it divides than are genes on plasmids.

- ARS plasmids (autonomously replicating sequences). Also YRp (yeast replicating plasmids). Have sequences from the yeast chromosome in them which allow them to be replicated as the cell divides.

All of these can be expression vectors to allow for the gene cloned into them to be used to make a protein. In addition, many yeast vectors are shuttle vectors. They have all the sequences needed to be effective cloning vectors in yeast cells, but they also have *E. coli* vector sequences in them. This allows the genetic engineer to shuttle the DNA between yeast cells (where he or she ultimately wants the recombinant DNA to reside) and *E. coli* cells (where it is more convenient to manipulate the DNA).

See also **Genetic code and protein synthesis, YACs**.

Yuk factor

A flippant term for the very real observation that the public, and indeed many scientists, judge the ethical acceptability of experimental procedures and biological manipulations in accordance with a scale of personal distaste. Thus the creation of the first cloned carrot in the 1960s was greeted with amusement in the press, while the creation of the first cloned frog in the early 1970s was treated with interest and some caution, and the possibility of mammalian cloning in the early 1980s was greeted with widespread horror. (This despite the fact that no adult mammal *had* been cloned.) Similarly, tests that rely on newts have less negative impact on public relations than those on rats, and rats are considered more acceptable than rabbits or dogs.

In general this reflects a concern for animals that look or behave more like human beings, or those that are kept as pets and hence attributed with human feelings. The ultimate public condemnation is therefore reserved for the potential scientific interference with human fetuses or children. This is a very real scale of values, and one that many scientists do not take seriously enough (hence they call it a 'yuk factor' rather than a 'value scale'). In public debate the yuk factor is sometimes a deciding one: much of the opposition to Monsanto's promotion of **BST** as a biopharmaceutical to boost the milk production of dairy cattle, is based not on arguments about farm economics but on the feeling that it must be horrible for the cow to be turned into nothing but a milk-producing machine.

Index

2-micron plasmid 337
2,4 dichlorophenoxyacetic acid 169
3' untranslated region 150
5' untranslated region 150
510(k) application 115
96-well plate 299
abzyme 66
  and coenzymes 91
ACDP, *see* Advisory Committee on
  Dangerous Pathogens, UK
acetic acid, volume of production 78
acetone, volume of production 78
acetylation of proteins 260
ACGM, *see* Advisory Committee on
  Genetic Manipulation, UK
acquaculture 22
acquired immune defficiency syndrome, *see*
  AIDS 7
actinomycetes producing secondary
  metabolites 292
activated sludge process 294
activation of support materials 4
acylases for biotransformation 59
ADA, adenosine deaminase 147
adapter oligonucleotides 276
adenosine deaminase deficiency and gene
  therapy 147
Adenovirus 1
ADEPT, *see* antibody-directed enzyme-
  prodrug therapy
adherent cells 11
adjuvant for immunization 184
advanced glycosylation end-products 160
adverse drug reactions and genetics 262
Advisory Committee on Dangerous
  Pathogens, UK 213
Advisory Committee on Genetic
  Modification, UK 213
aerobe 211
affinity chromatography 2, 85
affinity constant 28
affinity tag 4, 140
Affymax 101, 276
AFM, *see* atomic force microscopy
AFP tumour marker 324
agar 122
  in plant cell immobilization 252
agarose gel electrophoresis 142
AGE, advanced glycosylation end-products
  160
ageing and protein glycation 160

aggregates of cells in cell culture 210
Agrobacterium rhizogenes 167
Agrobacterium tumefaciens 5
  manipulated using recombination 173
AI, *see* artificial insemination
AIDS 7
  disease models 319
  fusion biopharmaceuticals and 139
  therapy based on glycobiology 161
  treatment protocol programme 322
airlift bioreactor 9, 206
alcohol, *see* ethanol
algae 23
  for biomass production 46
alginate
  for encapsulation 122
  in plant cell immobilization 252
allele specific oligonucleotide, *see* ASO 230,
  287
Allergan 315
alpha amylase 259
Alpha environmental 54
alpha-feto protein, AFP, tumour marker
  324
ALR airlift reactor 9
  see also airlift bioreactor
ALT, *see* autolymphocyte therapy
Altman, S. 287
American Type Culture Collection, ATCC
  96
Ames, B. 223
Ames test 223
Amgen
  patent dispute 240
  producing growth factors 166
amino acid modification in proteins 260
amino acids 10
  volume of production 78
  see also individual amino acides listed by
  name
ammonia as fermentation substrate 136
amphibolic pathways 292
amphipathic molecules 176
amplification of genes in expression systems
  132
ampometric enzyme electrodes 125
amylase, alpha amylase 162
  for food processing 137
anabolic pathways 292
anaemia 60
anaerobe 212

INDEX

anaerobic fermentation 39
anchorage dependent / independent cells 314
Anderson, W.F. 147
animal cell immobilization 11
ankolysing spondylitis 262
anther cultures 253
anti-foam agents 52, 135
anti-idiotypic antibodies 12
antibiotic 14, 291
  genoceuticals as 155
  hybrid molecules 175
  resistance genes 322
  resistance genes in Vector 328
  made by streptomyces 212
antibodies 15
  in affinity chromatography 3
  catalytic antibodies 66
  used in diagnostics 186
  as drugs 191
  as imaging agents 179
  enzyme stabilization using 129
  used in immunoassay 186
antibody
  immunization 184
  structure 17
  targetting in ADEPT 2
antibody-directed enzyme prodrug therapy,
  ADEPT 1
antifreeze proteins 95
antigen 5, 184
antimony purified using microbial mining
  209
antisense 19
  antiviral compounds 22
  Darwinian cloning of 100
  triple DNA and 323
antiviral compounds 21
  against new diseases 227
  genoceuticals and 155
  glycobiology and 161
apoenzyme 91
apoprotein 91
aptamer 100
  and triple DNA 323
archetype 224
Ares-Serono 321
Arabidopsis thaliana 156
  plant storage proteins 258
ARIS 188
Armour Pharmaceuticals 4
ARS, see autonomously replicating sequence
arsenic isolated using microbial mining
  209
artificial insemination 119
artificial sweeteners 23
ascites fluid, ascitic fluid 221
aseptic fermentation process 135
ASO 287
aspartame 23

as a synthetic peptides 245
  made using protease 264
aspartate in aspartame 23
Aspergillus 212
assymetric catalysis 81
astaxanthin 23
ATCC, see American Type Culture
  Collection
atomic force microscopy 289
atrazine 169
atropine 292
attached culture of mammalian cells 210
attenuated virus vaccines 331
attenuation of viruses 205
attitudes to biotechnology 224
  yuk factor 337
autoclave 301
Autographa californica 27
autologous lymphocyte therapy,
  autolymphocyte therapy 147
autolysis 69
autonomously replicating sequences, ARS
  337
auxins 121
auxostat 24
avidin 28, 58
  as an affinity tag 5
avirulence genes in plants 246
AZT, Retrovir 7, 21
azurocidin 15

B.thuringiensis, see Bacillus thuringiensis
BABS, see biosynthetic antibody binding
  sites
Bacillus subtilis 212
Bacillus thuringiensis 44, 51, 246
bactenecin 15
bacterial permeability increasing protein 15
bacteriophage 26
baculovirus 26, 44
banded iron formation 40
bands, in gel electrophoresis 143
barnacles 159
base in DNA 149, 144
basic fibroblast growth factor 226
batch fermentation 134
Baxter Healthcare 4, 61
BDNF, brain-derived neurotrophic factor
  226
Beckers muscular dystrophy 152
beer manufacture using enzymes 137
beta lactamase for biotransformation 59
beta-2-microglobulin, tumour marker 324
bFGF, basic fibroblast growth factor 226
BIAcore 234
binding 28
binding of proteins partly caused by sugars
  161

binding site of antibodies 18
bioaccumulation 29, 291
bioassay 29
biobleaching of wood 333
biocide 301
biocontrol 43
    using Baculovirus 27
    using trout for aquaculture 23
bioconversion 30
    and chiral synthesis 82
    in organic solvents 32
biocorrosion 37
biocosmetic 33
biodegradable materials 33
biodiversity 34
bioethics 35
biofilm 36
biofouling 37
biofuels 38
biogas 38, 39
    and sewage treatment 294
Biogen 8, 98
biohydrometallurgy 40, 208
bioinformatics 41
bioleaching, see leaching
biolistics 42
    and gene therapy 148
    and plant genetic engineering 255
biological containment 43
biological control 43, 51
biological oxygen demand, BOD 30
    detection using cell biosensor 182
    in sewage treatment 294
    and wood waste 334
biological response modifiers 45
biological transformation of chemicals 30
biologics 247
bioluminescence 207
biomass 46
    from aquaculture 23
    and SCP 290
biomaterials 47
    in biocosmetics 32
biomimetic 38
biomineralization 49
biomolecular engineering, recombinant
    DNA technology 276
biopesticide 51
biopharmaceutical 247
    altered by protein engineering 266
    fusion biopharmaceuticals 39
    glycosylation 163
    immunotherapeutic 190
    proteases as 264
    from ribozymes 287
    toxins as 315
    vaccines as 326
biopolymers 34
bioreactor 51

cleaning in place 134
and fermentation process 88
bioremediation 53
    and bioaccumulation 29
    and bioconversion 31
    and biodegradable materials 33
    and microbial mining 208
    of xenobiotics 335
biosensor 55
    electrochemical sensors 117
    enzyme electrode 127
    immobilized cell biosensor 181
    immunosensor 189
    ISFET 195
    optical biosensor 233
    thermal sensor 312
biosorption 56
    and bioaccumulation 29
    and biohydrometallurgy 40
    of sea water 291
biosynthetic antibody binding sites, BABS
    99
BioTechnica 228
biotechnology centers, USA 90, 308
Biotherapeutics, see biopharmaceuticals:
biotin 58
    binding of 28
    in biocosmetics 33
biotinylation 58
biotransformation 58, 59
    for AZT production 82
    and chiral synthesis 7
    to produce D amino acids 264
    proteases in 11
biphasic biocatalysis, reversed phase
    biocatalysis 285
bleaching of wood 333
blood disorders 60
blood as a fermentation substrate 127
blood products 61
blood–brain barrier 114
blot 62
    DNA probes in 109
    using hybridization 175
BOD, see biological oxygen demand
bone 50
bone marrow
    as gene therapy target 147
    transplant 234
booster in immunization 184
Borrelia burgdorfei 227
botulinus toxin 314
bovine growth hormone, BST 63
bovine serum albumin, BSA 184
    and monoclonal antibody 220
bovine somatotrophin, BST 63
    and the Yuk factor 338
bovine spongiform encephalopathy, BSE
    321

INDEX

BPI, *see* bacterial permeability increasing protein
Bradyrhizobium 228
brain-derived neurotrophic factor 226
Brassica napus, oilseed rape 258
breed, strain 302
BRMAC, Biological Response Modifiers Advisory Committee 45
bromoxynil 169
Brookhaven National Laboratory, USA 41, 92
Brookhaven Structures Database, see Protein Database 92
BSA, *see* bovine serum albumin
BSE, *see* bovine spongiform encephalopathy
BST, *see* bovine somatrophin
B.t. toxin, B.t.k.toxin 45, 51, 246
butanol, volume of production 78

CA 125, tumour marker 324
CA 19-9, tumour marker 324
cadmium purified using microbial mining 209
Caenorhabdis elegans genome project 156
caesium chloride centrifugation 76
caffeine 292
calcitonin
  as a peptide 244
  produced by genoceuticals 155
calcium phosphate used to get DNA into cells 316
Calgene 51, 246
callus
  generation 253
  regeneration from 297
Candida utilis 211
capillary zone electrophoresis, capillary electrophoresis 65
capsule 131
carageenan in plant cell immobilization 252
carbocyclic compounds as antiviral compounds 21
carcinoembryonic antigen, CEA, tumour marker 324
carcinogens detection by immobilized cell biosensor 182
cardiovascular disease, genetics of 262
carrier particles for cells 181
casein hydrolysate as fermentation substrate 136
cassette vector systems 329
castor bean oil 256
Castro case 108
catabolic pathways 292
catalytic antibodies 66
  and coenzymes 91
  selected with Darwinian cloning 100
catalytic RNA 287

and Darwinian cloning 101
CD antigens 98
CD4 8
CD4-IgG fusion protein 139
cDNA 68
  gene libraries 145
  sequences and patents 240
CDR, *see* complementarity determining region
CEA, tumour marker 324
Cech, T. 287
cell aggregates 210
cell compartment 132
cell culture
  in airlift fermenters 9
  tissue culture 313
cell disruption 69
cell fusion 70
  using electroporation 118
  to achieve cell immortalization 183
  to make monoclonal antibody 219
cell growth 71
  and fermentation process 134
cell line 74
  rights to 74
cell membrane 245
cell senescence 73
  and cell immortalization 183
cell wall 131
Celltech 191, 221
cellular adhesion molecules 179
cellular compartment 260
cellular hybridization, *see* cell fusion 175
cellulase
  as affinity tag 5
  for cell disruption 69
  as a glycosidase 162
cellulose as a fermentation substrate 136
Centocor 191
centrifugation 75
  for cell harvesting 168
Centromere 336
cereals as fermentation substrates 127
cerebrospinal fluid 114
Cetus 45, 98, 241
chain termination method for DNA sequencing 110
chaotropic agents 268
  to dissolve inclusion bodies 131
chaperones, chaperonins 77
cheese manufacture
  protease 264
  using enzymes 137
chelators in affinity chromatography 3
chemical design 218
chemicals made by biotechnology 77
chemiluminescence 207
chemostat 24
  fermentation process 134

chimera 78
chimeric antibody 80
  production 221
chinese hamster ovary cells, CHO cells 223
chiral chromatography 81
chiral intermediate 83
chiral synthesis 81
chirality 83
Chiron 8, 240, 323
chitin used in biosorption 57
chitinase 69
  pest resistance in plants 246
chitosan used in biosorption 57
Chlorella 46
chlorinated hydrocarbons 54
CHO cells, *see* chinese hamster ovary cells
chromatography 84
  as a purification method 273
chromosome 144
  chromosome jumping 332
  chromosome walking 332
Ciba Geigy 169
ciliary neurotrophic factor, CNTF 226
citric acid, volume of production 78
Clark electrode 117
class I / II proteins 257
clathrin 203
clean room 87
cleaning in place 88
Cline, M. 148
clone 89
  of animals 89
  of genes 89
  gene library 145
  of organisms 89
  of plants 297
cloning
  cells in monoclonal antibody generation 219
  embryos 119
  into a vector 277
  plants 252
cloning vectors, *see* vector
Clostridium acetobutylicum 211
clover 228
clubs 90
cluster differentiation antigens, *see* CD antigens 98
CMV, see cytomegalovirus
CNTF, *see* ciliary neurotrophic factor
coal 105
cobalt purified using microbial mining 209
cocci, miroorganisms 212
cocoa fat 256
coconut oil 256
coding region
  gene structure 144
  and genetic code 150
  codons 150

codon usage and gene synthesis 146
coenzyme, cofactor 90
Collaborative Research 138
collagen
  used in biocosmetic 33
  used as biomaterials 47
  encapsulation 122
  glycation of 160
collagenase for meat processing 264
colony blot 63, 282
colony stimulating factor 61, 98, 191
column chromatography, *see* chromatography
company names 225
competent bacteria 316
competitive assays, competition assay 187
complementarity determining region, CDR 17, 81, 99
complementarity of DNA strands 174
composite materials 33
composition of matter patents 239
computational chemistry 91
computerized tomography, CT 180
  and imaging agents 180
  and immunoconjugates 185
concentration 92
  in downstream processing 111
conjugation of bacteria 304
consortium
  in biofilms 37
  in fermentation process 135
constant region of antibody structure 18
containment 43
  containment laboratories 88
  containment level 249
  physical containment 249
continuous culture 34
contrast agent 185
control regions in gene 144
controlled pore glass, CPG 86
copper isolated by microbial mining 209
copy number
  in expression system 132
  and vectors 329
core particle of retroviruses 283
corn steep liquor as fermentation substrate 36
Cornell University 42
cosmid 328
  in gene library construction 145
  used for gene walking 332
cotton seed meal as fermentation substrate 129
countercurrent separation 92
cow/buffalo chimeric animal 78
cowpea tripsin inhibitor 246
cowpox 327
CPG, *see* controlled pore glass
crayfish 22

Creative Biomolecules 99
critical temperature / pressure 307
cross-flow filtration 94
crown ether 203
Crown Gall disease 5
cryopreservation 95
CSF, *see* cerebrospinal fluid, colony
  stimulating factors
CT scan, *see* computerized tomography
CTX, clinical trial exemption certificate 115
cultivar, strain 34, 302
culture collections 96
culture medium 135
cyclodextrins 97
  in affinity chromatography 3
  in biocosmetics 33
  in liquid membrane separations 203
cysteine 10
  and disulphide bond 105
  in wool 334
cystic fibrosis gene 285
cystine 105
cytokine 97
  blood products 61
  immunotherapeutic 190
cytomegalovirus
  AIDS 7
  new diseases 227

D-amino acids 11
Dab 99
  and antibody structure 17
  and chimeric antibody 81
Darwinian cloning 99, 100
  to produce ribozyme 287
death phase of cell growth 73
debranching
  enzyme 162
  and polysaccharide processing 260
debridement 264
defensin 15
dehydrogenase enzymes 38
delayed fluorescence, DELFIA 102
deliberate release 104
  regulation of organism release 279
denaturation of proteins 268
  used in purification 272
density gradient centrifugation 76
Department of Trade and Industry, UK,
  DTI 90
depression, genetic predisposition to 262
desulphurization 105, 208
detergents, proteases in 263
determined cells 300
dextrins 136
di-deoxy method of DNA sequencing 110
diabetes
  genetic predisposition to 262

models in mice 320
dichlorophenoxyacetic acid 169
differentiation 300
dilution of antibodies 185
dilution rate in chemostat 24
dimethylsulphoxide, DMSO 95
  and cryopreservation 95
  hydrophobicity 176
diphtheria toxin 192
diploidization 253
disengagement of gas in airlift fermenter 9
dissociation constant 28
disulphide bond 105
  and antibody structure 18
  and protein stability 268
DMSO, *see* dimethylsulphoxide
DNA 144
  amplification 106
  fingerprinting 107
  hybridization 174
  ligase 276
  sequences, patents for 240
  sequencing 110, 142
  synthesiser 146, 230
DNA Plant Technologies 246
DNA polymerase as a reagent 276
DNA probe 108
  diagnosing AIDS 7
  for DNA fingerprinting 107
  and genetic information 154
DOE funding for genome project 156
domain in antibody structure 17, 99
dominant genes 152
dot blot 63
doubling time 73
dough conditioning using enzymes 137
Dow Chemical 138
downstream processing 111
Draaijer, J. 299
Drosophila 156
drug delivery 113
  and liposomes 202
drug development pathway 115, 116
drug discovery
  using computational chemistry 92
  using receptor binding screening 275
DTI, *see* Department of Trade and
  Industry, UK
Du Pont 37, 42
Duchenne muscular dystrophy 152
dump leaching 200
Dutch elm disease 44
dye front 143
dye–ligand chromatography 3

E.coli see Escherichia coli
EC cells 79
EC number 124

ECACC, *see* European collection of animal cell cultures
ectopic genes in transgenic plants 318
Edman degradation 267
EFB, *see* European federation for biotechnology
effects of ageing reversed by human growth hormone 174
EGF, *see* epidermal growth factor
ELAM, see endothelial-lymphocyte adhesion molecule
electroblot 62
electrochemical sensors 117
electron spin resonance, ESR 180
electrophoresis in DNA sequencing 111
electroporation 118
elicitors
    in plant cell culture 251
    of secondary metabolites 292
ELISA 186
ELM, *see* emulsion liquid membrane 204
elution profile 84
EMA, tumour marker 324
EMBL DNA data library 41
embryo splitting 119
embryo technology 119, 78
embryogenesis in plants 121
    and micropropagation 214
embryonal carcinoma cells, EC cells 79, 317
Eminase 61
EMIT 189
emulsion liquid membrane, ELM 204
enablement in patents 240
enantiomer 83
enantiomeric excess 59
encapsulation 122
    in drug delivery 113
    using liposome 202
endothelial cells 179
endothelial-lymphocyte adhesion molecules, ELAM 161
ENFET 196
enhancer 145
enrichment of microorganisms 305
entamophagous fungi 44
entropy of reaction 67
env gene of retrovirus 283
environmental biotechnology 123
    desulphurization 105
Environmental protection agency, USA 279, 281
Environmental protection law, UK 279
EnzFET 196
enzyme 129
    amounts produced 129
    conjugates with antibodies 186
    enzymes used in biotransformation 58
    mechanisms 126
    nomenclature 124

production by fermentation 127
specificity altered by protein engineering 266
stabilization using antibodies 128
used in affinity chromatography 3
Enzyme commission 125
enzyme electrode 125
    electrochemical sensors 117
    enzyme immobilization on 125
enzyme immunoassays 207
enzyme-linked immunosorbant assay, see ELISA 186
eosinophilia-myalgia syndrome, EMS 10
EPA, *see* Environmental protection agency, USA
ephedrine 30,83
epidermal growth factor 165
episome 259, 328
epithelial cells as targets for drug delivery 114
epithelial membrane antigen, tumour marker 324
epitope 15
epizooic 44
EPO, *see* erythropoietin, European patent office
Eppendorf 299
equilibrium centrifugation 75
erb-B2 oncogene 231
erythropoietin 60, 165
    and patent disputes 240
    and sports misuse 299
Escherichia coli 156, 213
ESR, *see* electron spin resonance
essential amino acids 10
EST, *see* expressed sequence tag
esterase 59
ethanol
    produced by bioconversion 30
    as a biofuel 38
    volume of production 78
ethylene oxide 301
eukaryote 212
European collection of animal cell cultures, ECACC 96
European Commission
    directives on organism release 280
    funding for genome project 156
European federation for biotechnology, EFB 211
European Molecular Biology Laboratory, EMBL 41
European patent office, EPO 239
evanescent wave sensors 233
Exactech 56
exogenous enzymes in food processing 137
exogenous genes in transgenic animals 318
exon 150
explant 253
expressed sequence tags, EST 68

expression compartment 130
expression system 132
   adenovirus as an 1
   Baculovirus 26
   cDNA 68
   maxicell 208
   plant storage proteins as 258
   protein stability 269
   secretion 293
   transgenic animals as 318
   vaccinia virus as 327
expression vector 329
extreme thermophile 312
   biological response modifiers 45
   gene therapy-regulation 149
   food processing using enzymes 138

Fab, Fab' 17
FAB-MS, fast atom bombardment MS for
   protein sequencing 267
Factor IX 60, 61
Factor VII 60
Factor VIII 4, 60
   and patent disputes 240
facultative aerobes, anaerobes 212
FAD, see foavine adenine dinuclsotide
farm animal engineering 319
fast protein liquid chromatography, FPLC
   273
fast track drug development 8
Fc 17
FCS, see fetal calf serum
FDA, see Food and Drug Administration
fed batch fermentation 134
fermentation process 134
   enzyme production by fermentation 127
   fermentation medium 135
   fermentation substrate 127
   gas transfer 141
   harvesting 168
fermentation to achieve chiral synthesis 82
fermenters, bioreactor 310, 52
Ferrobacillus 209
ferrocenes 125
fetal calf serum 314
fetal research 36
FGF, see fibroblast growth factor
FIA 188
fibre optic chemical sensor 234
fibrin 159
fibroblast growth factor, FGF 166
fibroin 47
fibronectin 33
field effect transistor 195
filamentous organisms 212
filter bioreactor 181
filters using 170
filtration

   for harvesting cells 168
   for sterilization 302
financial support 308
fish culture 22
fish meal as fermentation substrate 127
fixing nitrogen 228
FLAG peptide 5
flavine adenine dinucleotide, FAD 38, 91
flavour compounds
   made in hairy cell root culture 167
   secondary metabolites 292
flavour modification by food processing
   using enzymes 137
flax oil 256
floatation 94
flocculation
   for cell harvesting 168
   for concentration 94
fluidized bed 181
fluorescence, DELFIA 102
fluorescence immunoassay, FIA 188
FOCS, see fibre optic chemical sensor
Food and Drug Administration, FDA 280
   biological response modifiers 45
   gene therapy-regulation 149
   food processing using enzymes 138
food nutritional value 257
food processing using enzymes 137
formulation of drugs 113
formulation to increase protein stability 268
fos oncogene 231
FPLC, see fast protein liquid
   chromatography
fragrance compounds
   from hairy cell root culture 167
   secondary metabolites 292
free zone electrophoresis 65
freeze-drying 138
   and cryopreservation 95
freeze–thaw 69
freezing for cryopreservation 95
French press 69
Freunds complete adjuvant 184
fructose 158
fusion biopharmaceuticals 139
fusion of cells 70
fusion phage 100
fusion protein 140
   containing affinity tag 4
   for peptide synthesis 243
   substrate chanelling 307

G-CSF 61
gag gene of retrovirus 283
gamete intrafallopian transfer, GIFT 119
gamma globulin, immunoglobulin 61
gamma-carboxyglutamate 262
gamma radiation for sterilization 302

gas as a fermentation substrate 136
gas transfer 141
gasohol 38
geep 78
gel chromatography 84
gel electrophoresis 141
gel filtration 273
gelatin in plant cell immobilization 252
GenBank 41
gene 144
  amplification in expression system 132
  clone 145
  cloning 89
  cloning using PCR 242
  jumping 332
  library 145
  machine, DNA synthesiser 146, 230
  synthesis 146
  walking 332
gene targetting using homologous
  recombination 172
gene therapy 147
  biolistics 42
  genetic engineering 153
  and genoceuticals 155
  and immunotherapeutics 192
  regulation of 148
Genentech 8, 61, 98, 139, 173, 240, 306
generally regarded as safe, GRAS 165
genetic antibiotics 155
genetic code 149
genetic component to disease 262
genetic disease diagnosis 152
  in embryos 120
  using PCR 242
genetic engineering 153
  plants 254
  strain development 304
genetic fingerprinting, DNA fingerprinting
  107
genetic information 36, 154
Genetic Information Act, USA 154
genetic manipulation, genetic engineering
  153
genetic map 156
genetic predisposition to disease 262
genetically manipulated organism, GMO
  104
  recombinant DNA technology 277
  regulation of release 149
Genetics Institute 75, 166, 240
Genex 99,159
genoceuticals 155
genome project 156
  and bioinformatics 41
  DNA sequencing 110
  genetic information 154
  predisposition analysis 263
  YACs for 335

genomic gene library 145
genotype
  disease diagnosis 152
  reverse genetics 284
GenPharm 133
Genta Inc. 20
Genzyme 318
germ-line gene therapy 147
GIFT, see gamete intrafallopian transfer
Gilson, pipetter 299
GLP, see good laboratory practice
glucagon 244
glucanase 159
glucose biosensor 56
  enzyme electrode 126
glucose isomerase 158
  coenzyme for 162
  glycosidases 260
  polysaccharide processing 91
glue 159
glutamate 10
  hydrophobicity of 176
  volume of production 78
glycation 160
glycerol for cryopreservation 95
glycobiology 160
glycoform 163
  of tumour markers 325
glycolipid 161, 163
Glycomed 161
glycoprotein 160, 162
  ICAM as a 179
glycosidases 162
  for food processing 137
glycosylation 162
  and protein pharmacokinetics 248
  in yeast cloning systems 336
  post-translational modification 260
glycosyltransferase 161
glycotechnology 161
glyphosate 169
glypiation 161
GM-CSF 61, 139
GMMO, genetically manipulated
  microorganism 104
GMO, see genetically manipulated organism
GMP, see good manufacturing practice,
  good microbiological practice
gold 164
good laboratory practice, GLP 157
  drug development programes 115
good manufacturing practice, GMP 157
good microbiological practice, GMP 158
gp120, gp160 8
grade of metal ores 199
gradient chromatography 86
gram positive / negative microorganism
  212
GRAS, see generally regarded as safe

growth factor 165
  neurotrophic factors 226
growth hormone
  in wool production 334
  sports 298
growth rate of cells 73
guest molecule in cyclodextrins 97

haematopoiesis 60
  and growth factors 166
  stem cell growth factor 301
haemoglobin 60
haemophilia 60
  genetic disease diagnosis 152
hairy cell leukemia 75
hairy cell root culture 167
halotolerance 237
HAMA, see human anti-mouse antibody
  response
haploid plants 253
hapten 220
harvesting 168
  and fermentation process 134
  as a purification method 270
  in downstream processing 111
HCFA, see Healthcare Financing
  Administration, USA
hCG, Tumour marker 324
HCGF, haematopoietic cell growth factor
  166
head space in bioreactor 52
Healthcare Financing Administration, USA
  281
heap leaching 200
heat denaturation 272
heavy chain in antibody structure 17
HeLa cells 75
hepatitis viruses 227
herbicides 168
herpes simplex 227
heuristics in computational chemistry
HFCS, see high fructose corn syrup
hGH, see growth hormone 173
high fructose corn syrup, HFCS 59, 158
high pressure liquid chromatography, HPLC
  86, 273
HIV 7, 284
HIVER theory of AIDS 7
Hoecst 169
Hoffman La Roche 242, 306
hollow fibre 170
  for animal cell immobilization 11
  bioreactor 181
  improved gas transfer 141
  for plant cell immobilization 252
homogenous immunoassay 188
homologous genes 109
homologous recombination 171

in engineering Baculovirus 27
in transgenic disease models 320
hormones, anti idiotypic antibodies as
  mimics 13
horse radish peroxidase, HRP 186
HPLC, see high presure liquid
  chromatography
HSA, see human serum albumin
HTLV virus family of retroviruses 284
Hu-SCID mouse 320
HUGO 156
human anti-mouse antibody response
  HAMA 80
human chorionic gonadotrophin 324
Human Genome Organisation, HUGO
  156
human genome project, 156
  and bioethics 35
  DNA sequencing 110
  see also genome project
human serum albumin, HSA 61
humanized antibody 80, 193, 221
humanized, humanoid mice 319
humus 296
hyaluronic acid 33
hybrid antibiotics 14
hybrid crops 256
hybridization 174
  of DNA probe 303
  of organisms 108
hydrodesulphurization 105
hydrogen 38
hydrolysis of plant oils 256
hydrophilicity 76
hydrophobic catalysis, organic phase
  catalysis 236
hydrophobic chromatography 85, 176
  as a purification method 273
hydrophobicity 176
  and protein stability 268
  hydrophobicity plot 177
hydroxyproline 261
hypersensitivity to drugs 262
hypertension, genetic predisposition to
  262
hyphus, hyphae 212
hypoallergenic 33
hypoglycaemia 244

ICAM, see intracellular adhesion molecules
identifying useful microorganisms 304
IL, see interleukin
ILM, see immobilized liquid membrane
imaging agent 185, 179
immobilized cell bioreactors 180
  for monoclonal antibody production 221
immobilized cell biosensor 181
immobilized liquid membrane, ILM 203

immortalization of cells 74, 183, 316
immune function modelled in transgenic
    disease models 320
immune response 12
immune system
    and biological response modifiers 45
    disorders of 262
    immunization 184
    and monoclonal antibody generation 219
    as target for antiviral compounds 21
Immunex 98
immunoassay 186
    using immunoconjugates as labels 186
    using luminescence 207
immunoconjugate 185
immunodiagnostic 186
ImmunoGen 192
immunogenicity of vaccine 205, 327
immunoglobulin, antibody 15
immunosensor 189
immunotherapeutic 190, 191
immunotoxin 192
    targetted drug delivery 311
    toxins for 315
impeller in bioreactor 52
in situ sterilization of bioreactors 88
in utero diagnosis 120
in vitro fertilization 119
in vitro 193
in vivo 193
inclusion bodies 130, 133
incremental selection of strains 303
incubation 134
inducer 194
    of enzyme production by fermentation
    128
induction, of gene 194
    in expression system 132
ingenieur biomoleculaire 278
inherited disease diagnosis 152
initiation in embryogenesis in plants 121
inoculation 194
    of soil with nitrogen fixing bacteria 228
integration of DNA into chromosomes 172
interferon 98, 191
    used in animal breeding 21
    as antiviral compounds 98
interleukin 98, 191
    and gene therapy 147
    Interleukin-2 45
    Interleukin-3 139
interspecific hybrid 175
intracellular adhesion molecules, ICAM 179
intron 150
    spliced out by yeast 336
inverse PCR 242
invertase 158
Investigational New Drug application (IND)
    115

in polysaccharide processing 260
ion-exchange methods of concentration 92
    as a purification method 273
ion-exchange membrane 204
ion-sensitive field effect transistor, ISFET
    195
ionomers 125
iontophoresis 114
ISFET, see ion-sensitive field effect
    transistor
ISIS Pharmaceuticals 20
isoamylase 162
isoelectric focussing 142
isoelectric point of protein 142
    use of purification 272
IVF, see in vitro fertilization

Jeffreys, Prof. A. 107
jojoba wax 256
Jung, K. 224

K_a, affinity constant 28
Kabi Pharmacia 298
K_d, dissociation constant 28
keratin in wool 334
kevlar 47
keyhole limpit haemocyanin, KLH 184, 220
kilobase 144
klenow polymerase 276
Krutzfelt–Jacob syndrome 321
Kuhn, T. 216
kuru 321

labels 186
    biotin as 58
    immunoconjugates 186
    labelling DNA 109, 277
    using luminescence 207
laboratory safety classification 211
lactate dehydrogenase 266
lactoferrin 15
lactose as stabilizer for freeze-drying 138
lag phase in cell growth 71
lambda bacteriophage 26
    as cloning vectors 328
Langmuir–Blodgett film 197, 199
    as liquid membranes 204
    in molecular computing 217
latex agglutination assay 187
leaching 199
    gold 209
    and microbial mining 164
leader sequence 222
lectins 3
leghaemoglobin 228
legumes 228

lesquerella oil 256
leukaemia 60, 165
ligand 2
ligase chain reaction 106
light chain in antibody structure 17
lignin
  in biosorption 57
  in wood 333
ligninase 333
LINK scheme 90
linkage analysis 152
linked marker 287, 332
linker oligonucleotides 278
lipase 59, 200
  for food processing 137
lipid bilayer:
  Langmuir–Blodgett film 197
  liposome 201
lipid 200, 201
lipid membranes
  hydrophobicity 177
  Langmuir–Blodgett film 197
lipophilic 176
liposome 201
  in biocosmetics 33
  and drug delivery 113
  and gene therapy 148
  glycosylation 163
liquefaction 259
liquid membrane 197, 204
  liquid membrane separations 203
liquid nitrogen storage 95
liquid–liquid separation 270
live vaccine 204
  adenovirus 1
  vaccinia virus 327
  viral vaccines 330
lobsters 22
log phase of cell growth 71
logical reasoning 194
long terminal repeat, LTR 283
loop bioreactor, fermentor 206
luciferase
  immobilized cell biosensor 182
  luminescence 207
luminescence 207
luminescent bacteria 182
lupin 228
lyme disease 227
lymphocytes 16
  used as immunotherapeutic 192
  and inflammation 179
  monoclonal antibodies produce from 219
lyophilization 138
lysine 10
  hydrophobicity 76
  volume of production 78
lysozyme
  as an antibiotic 15

used in cell disruption 69
lytic bacteriophage 26

m13 26
  used in site-directed mutagenesis 295
mad cow disease 321
magnetic resonance, nuclear magnetic
  resonance 185
magnetite 50
maize 6
malate dehydrogenase 266
male sterility 257
malt extract 136
mammalian cell culture 171
  in airlift fermenter 9
  using microcarriers 209
MAP, see multiple antigen peptides
mariculture 22
mass sensors 189
mass spectrometry, MS, for protein
  sequencing 267
maturation in embryogenesis in plants 121
Maxam and Gilbert method 110
maxicell 208
meat tenderization using enzymes 137, 264
mediator compounds for enzyme electrodes
  125
megabase 144
membrane bioreactor 181
membranes, see cell membrane
Merrifield synthesis 244
mesh bioreactor 181
Messing, J. 26
metal affinity chromatography 3
metallothionein 57
metals extracted from sea water 291
methane 38, 39
Methanococcus 290
methanol 38
  as a fermentation substrate 136
methionine 10
methyl viologen 297
methylation of DNA 276
micro-centrifuge 300
microbeads 12
microbial leaching 199
  of gold 164
microbial mining 208
  and bioconversion 31
  and biomineralization 49
  biosorption 57
  leaching 291
  from sea water 199
microbial transformation of chemicals,
  bioconversion 30
microcarrier 209
  for animal cell immobilization 12
microchip sensors 189

microencapsulation 122
microinjection
  in plant genetic engineering 254
  transgenic animals 317
micromechanics 217
micronucleus test 223
microorganism safety
  classification 210
microorganisms 212
microparticle gun system 42
microparticles
  immunoassay 188
  latex assay 187
micropropagation of plants 214
  and embryogenesis 121
microtitre plate 299
  in immunoassay 188
Miles Inc 61
milk 133
  and BST 64
Milstein, C. 219
minicell 208
minimal recognition unit, MRU 99
minisatellite DNA 107
MIPS 41
miscelle 202
mismatches in DNA 174
MITI 90
mobile phase 84
molecular biology 215
molecular computing 216
  Langmuir–Blodgett film 199
molecular graphics 217
  and computational chemistry 91
molecular hybrid 175
molecular imprinting 39
molecular lego 215
molecular modelling 218
mollases 127, 136
molybdenum purified using microbial
  mining 209
monoclonal antibody 16, 219
  made in airlift fermenter 71
  cell fusion and 9
  production 221
  production in hollow fibres 171
monoculture 35
monosodium glutamate 10
Monsanto 44, 63, 169, 246, 338
Moore, J. 75
motifs in DNA or protein 222
mouth feel of food 162
MRU, see minimal recognition unit
mucopolysaccharides 37
mud, in oil extraction 209
Mullis, K. 241
multiple antigen peptides, MAP 327
multiwell plate 299
mussels 159

mutagenicity tests 222
mutations created by site-directed
  mutagenesis 295
myc oncogene 231
myc-y-mouse 232
mycoinsecticides 44
mycoplasma 7
myeloma cell line 220
mythogenesis 223
myxomatosis 43

N-formyl methionine 260
N-terminal methionine 260
N-terminus 151
NAD, see nicotinamide adenine dinucleotide
NADP 91
names for biotechnology companies 225
nanotechnology 217
  scanning tunnelling microscopy 289
nasal drug delivery 114
NASBA, see nucleic acid sequence-
  dependent amplification
National Biotechnology Policy Board, USA
  280
National Cancer Institute 149
National Center for Biotechnology
  Information, NCBI 41
natural gas 39
natural killer cells, NK cells 192
NBPB, see National Biotechnology Policy
  Board
NCBI, see National center for
  Biotechnology Information
NDA, new drug application 116
negative pressure containment 249
nerve growth factor, NGF 226
networks of antibodies 12
neu oncogene 231
neurone-specific enolase, NSE, tumour
  marker 324
neurotrophic factor 226
neurotropin 3, NT-3 226
new diseases 227
NGF, see nerve growth factor
nickel purified using microbial mining 209
nicotinamide adenine dinucleotide, NAD
  38, 307
  and affinity chromatography 3
  biomimetics for 38
nif genes 228
NIH
  funding for genome project 156
  regulation of gene therapy 149
nitrogen fixation 228
NK cells, see natural killer cells
NMR, see nuclear magretic resonance
nod genes 228
nodules 228

non-aqueous phase catalysis 236
non-polar molecules 176
nonsense mutation 151
nopaline 6
northern blot 63
NSE, tumour marker 324
nuclear magnetic resonance, NMR 180
  imaging agents for scans 180
  immunoconjugates scans 185
  structures for proteins 41
nuclear polyhedrosis virus 27
nuclear transplant 119
nuclease
  in cell disruption 323
  and triple DNA 69
nucleic acid sequence-dependent
  amplification, NASBA 106
nucleosides, production volume of 78
nutrasweet 23

O'Farrel gels 142
obligate aerobes, anaerobes 212
octopine 6
off label use 116
Office of Science and Technology Policy,
  USA, OSTP 280
Office of Technology Assessment, USA,
  OTA 308
oil
  from biofuels 39
  bioremediation of spills 54
  desulphurization of 105
  as fermentation substrate 136
  production from plants 296
  recovery using microbial mining 209
oilseed rape Brassica napus 258
oligonucleitode synthesiser 230
oligonucleotides 230
  as DNA probes 110
  in gene synthesis 146
  used for site-directed mutagenesis 295
oncogene 231
  antisense therapeutics to 19
  immortalization and 324
  as tumour marker 183
oncomouse 232, 231
operator 145
operon 144
opsonization 15
optical biosensor 233
Opticore 12
oral drug delivery 114
ore grade 199
organ culture 234
organ transplant 234
organic metals 126
organic phase catalysis 236
  lipases in 244

and peptide synthesis 200
organogenesis in plant cell culture 121
origin of replication 328
orphan drug act 237
osmoprotectant 238
osmotic shock 69
osmotolerance 237
osteoporosis
  and genoceuticals 155
  and peptide hormones 244
OSTP, see Office of Science and Technology
  Policy
OTA, see Office of Technology Assessment
oversight 238
Oxford Glycosystems 161
oxidase enzymes 117
oxygen electrode 117
oxygen 52
oxygen supply in fermentation 141

p24, HIV 8
packaging kits 277
PAGE, see polyacrylamide
palm
  oil 256
  tree propagation 298
PAP, tumour marker 324
paradigm of molecular biology 216
particle gun 42
Patent and Trademark Office, USA, PTO
  240
patents 239
  bioethics 36
  for cDNA 68
  organism deposit requirements 96
Pauling, L. 66
pBR series vectors 328
PCBs, bioremediation of 54
PCL, poly-caprolactone 34
PCR, see polymerase chain reaction
PDGF, see platelet-derived growth factor
pectin processing 260
PEG, see polyethylene glycol
penicillin made by biotransformation 59
Penicillium 213
peptidase for biotransformation 59
peptide mimetics 38
peptide production from fusion protein
  140
peptide synthesis 243
  using organic phase catalysis 236
peptide synthesizer 244
peptides 244
peptone as a fermentation substrate 136
Percoll 76
periplasmic space 131
permeabilization of cells 245
pertussis toxin 314

pest control 44
pest resistance in plants 246
PET scans, *see* positron emission
    tomogaphy
Petunia 51
PFGE, *see* pulse field gel electrophoresis
pH electrode as basis of electrochemical
    sensors 117
phage group 215
pharmaceutical proteins 247
Pharmaceutical Proteins Ltd. 318
Pharmacia 102, 234
pharmacokinetics 248
    and drug development programmes
    115
pharming 133, 318
Phase I, II, III trials 115
Phase IV 115
PHB, poly-hydroxybutyrate 34
phenotype 284
phenylalanine 10
    in aspartame 23
phosphoinothricin, PPT 169
phospholipids 176
phosphorothioate DNA analogues 20
photosynthesis 297
    in loop bioreactor 206
    disrupted by herbicides 169
photosystems I / II 297
PHV, *see* polyhydrovaleric acid
physical containment 249
pig genetic engineering 104, 319
pilot plant 52
PIR 41
pisciculture 22
pituitary gland 173
pK_i 272
PLA, product licence application 116
plant cell culture 251
    hairy cell root culture 167
plant cell immobilization 252
plant cloning 252
    somaclonal variation and 297
plant embryogenesis 121
plant genetic engineering 254
Plant Genetic Systems 246, 258
plant oils 256
plant sterility 256
plant storage proteins 257
plant tissue culture 167, 214
plaque 26
plaque lift 63, 282
plasma membrane, cell membrane 245
    broken in cell disruption 69
plasmid 258
plasmolysis 69
platelet-derived growth factor, PDGF
    166
plug flow reactor 53

PMA, pre-marketing approval 116
pol gene 283
polar molecules 176
poly-A tail 150
polyacrylamide
    for gel electrophoresis 142
    in plant cell immobilization 252
polycaprolactone 34
polyclonal antibody 16, 219
polydextrose 47
polyethylene glycol, PEG 70
    used for cell fusion 70
    to get DNA into cells 316
    used for purification 272
polygalacturonidase 20
polyhydrovaleric acid, PHV 34
polyhydroxybutyrate 34, 47
polylinker 329
polymer precipitation 272
polymerase chain reaction, PCR 241
    and diagnosis of AIDS 7
    DNA fingerprinting 107
    patent dispute 240
polyprotein vaccines 326
polysaccharide processing 259
polyurethane foam for plant cell
    immobilization 252
pond bioreactor 53, 141
Porton International 315
positron emission tomography, PET
    180
    imaging agents and 180
    immunoconjugates and 185
post-marketing surveillance 116
post-translational modification 260
    glycosylation 163
potato cloning 298
potentiometric enzyme electrode 125
PPT, *see* phosphoinothricin
pre-embryo 120
preclinical research 115
predisposition analysis 262
Presidents Office of Science and Technology
    Policy 280
pressure cycle bioreactor 206
primary immune response 184
primers 230
    PCR 241
Prion 321
proalcool 38
process patent 240
prodrug
    in ADEPT 113
    and drug delivery 1
    in targetted drug delivery 311
prokaryote 212
promoter 144
    and induction 194
prostate acid phosphatase, PAP 324

protease 263
  amount made 263
  for biotransformation 59
  for food processing 137
  in organic phase catalysis 236
  and peptide synthesis
protected fermentation process 135
protected N-terminus of proteins 260
protein biomass, SCP 290
protein crystallization 265
Protein Database 92, 41
protein engineering 265
Protein Engineering Research Institute,
  Japan 90
protein folding
  and chaperones 77
  computational chemistry 91
Protein Identification Resource, PIR 41
protein production using secretion 293
protein refolding from inclusion bodies
  131
protein sequence 266
  from gene sequence 267
protein stability 268
  improved by protein engineering 266
  in thermophiles 313
protein structure
  formed by chaperones 77
  computational chemistry 91
  and crystallography 265
  and rational drug design 274
  remodelling 114
  stabilized by disulphide bonds 105
  stabilized using antibodies 128
protein synthesis in cells 149
protein three dimensional structure 265
protoplast-2 69
  transformation of 255
Protos 276
provirus 282
pruteen 52, 290
pseudoaffinity chromatography 3
pseudomonas 213
  pseudomonas oleovorans 31
  pseudomonas toxin 192
pseudopregnant animals 78
PTO, see Patents and Trademarks Office,
  USA
public concern / information about
  biotechnology
  mythogenesis 223
  yuk factor 338
pullulanase 62, 259
pulmonary drug delivery 114
pulping wood 333
pulse field gel electrophoresis, PFGE 142
purification 57
  in downstream processing 111
  of metals 209

of proteins, and protein stability 269
  large scale 270
  small scale 273
  of thermostable proteins 313
purification tag, affinity tag 4
pyrogen as contaminant in pharmaceutical
  proteins 247
pyrolysis to make oil 39

Q-beta, Q beta replicase 106

rabies vaccine 331
race, of animal 302
racemate 82
radio-immunoassay, RIA 188
random amplification of polymorphic DNA,
  RAPD 107
rapeseed oil 256
ras oncogene 231
rational drug design 274
  molecular graphics and 218
  molecular modelling and 218
reading frame 151
receptor proteins 275
  binding for drug screening 275
Receptortech 276
recessive genes 152
recombinant DNA technology 277, 328
  bits and kits 276
recombinant proteins and protein stability
  269
recombination 171
redox reactions, reduction/oxidation
  reactions 40
regeneration of cells from protoplast
  270
regiospecific biotransformation 59
regulation 279
  deliberate release 165
  gene therapy 104
  genetic engineering 148
  organism release 279
regulatory authorities, US 280
rennin 137, 264
replica plate 281
replicon 328
represser 194
repressible gene, repression 194
repressors of enzyme production 127
resin, ion exchange 92
resin removal from wood 333
resistance to herbicides 168
resolution 59, 82
resonant frequency sensors 189
restorer genes 257
restriction enzyme 277, 286

restriction fragment length polymorphism,
  RFLP 286
  and genome project 156
  for gene walking 332
restriction map, site 277
  in plasmids 259
retrotransposons 284, 322
Retrovir, AZT 7, 21
retrovirus 282
  vectors for gene therapy 147
  vectors for transgenic animals 317
reverse genetics 284
  and genetic disease diagnosis 152
reverse miscelle 202
reverse osmosis 92
reverse transcriptase 68, 277
reversed phase biocatalysis 285
RFLP, see restriction fragment length
  polymorphism
rhizobium 228
RIA, see radio-immunoassay
ribozyme 287
  and Darwinian cloning 101
  as antiviral compound 22
rice
  containment of transgenic plants 43
  engineering by A.tumefaciens 6
  nitrogen fixation 228
ricin 314
  immunotoxins 192
RNA 144
RNA polymerase 144, 277
RNase H 19
root nodules 228
rotor 76
Roundup 169
rumen bacteria 335
runaway replication plasmids 132, 329
rushton turbine 310
ruthenium compounds to capture solar
  energy 297

saccharification 259
Saccharomyces cereviseae 213, 336
safety classification of microorganisms 210
salmon 22
salt precipitation 270
salt tolerance 237
Sandoz 75, 246
sandwich assay 187
Sanger method 110
saturated fat 200
SAW sensors 189
SCA, single chain antibody 81, 99
scale-up 288
scanning tunnelling microscopy, STM 289
SCBU, see special care baby unit
SCF, see stem cell factor

SCF, see supercritical fluid
schizophrenia 262
SCID, see severe combined
  immunodefficiency syndrome
Science and Engineering Research Council,
  UK, SERC 90
SCP, see single cell protein
Scripps clinic 240
SDS, see sodium dodecyl sulphate
sea water 291
secondary immune response 184
secondary metabolite 291
  and fermentation process 135
  hairy cell root culture 167
  in plant cell culture 251
secretion 293
  in Bacillus subtilis 212
  and expression compartment 132
  and motifs in protein 222
secretion vector 133, 293
sedimentation 92
Seeman, N. 323
selection of strains 303
self vs non-self 16
senescence of cells
  and cell growth 73
  and cell immortalization 183
separation
  in downstream processing 141
  by gel electrophoresis 111
  using maxicells 208
Sephadex 86
Sepharose 86
sepsis 191
SERC, see Science and Engineering
  Research Council, UK
severe combined immunodeficiency
  syndrome, SCID 147
sewage
  biofuels 38
  biogas 39
  treatment 293
  treatment as bioremediation 53
shear forces 9
sheep 334
  sheep/goat chimera, geep 78
shell space in hollow fibre 170
shigella toxin 192
shotgun cloning, sequencing 110
Showa Drenko KK 10
shrimp 22
shuttle vector 329
  in yeast cloning 337
sickle cell disease 60
signal peptide
  motifs in protein 222
  and post-translational modification 260
  secretion 293
single cell protein, SCP 46, 290

single chain antibody, SCA 81, 99
single-sided PCR 242
site directed mutagenesis 127, 295
size exclusion chromatography 84
size standards 143
skin 234
slope leaching 200
slot blot 63
sludge 294
smallpox 326
SmithKline Beecham 61
sodium dodcyl sulphate, SDS 142
soil
    amelioration 296
    bioremediation 304
    as source for microorganisms 53
solar energy 296
solid bed bioreactor 181
solvent catalysis, organic phase catalysis
    236
somaclonal variation 214, 297
    and plant cloning 253
somatic cell embryogenesis 121
somatic cell gene therapy 147
somatic gene therapy 42
SOP, see standard operating procedure
SOS chromotest 223
Southern blot 62
Southern, Prof. E. 62
soy flour 127
soy protein 136
space shuttle 265
sparger 9, 52, 310
special care baby unit, SCBU 120
species hybrids 175
sphingosine 33
spiders in tank bioreactors 310
spirulina 46
splenectomy to make monoclonal antibody
    219
splitting cells during cell growth 73
sports and biotechnology 298
SPR, see surface plasmon resonance
standard laboratory equipment 299
standard operating procedure, SOP 157
starch
    and biodegradable materials 33
    and biofuels 38
    as a fermentation substrate 127
    processing 259
State vs Castro 108
stationary phase of cell growth 71
stationary phase chromatography 84
stem cell factor, SCF 166, 300
stem cells 300
stereospecific biotransformation 59
sterile fermentation process 135
sterilization 301
    in fermentation process 134

steroids
    made by bioconversion 31
    made by biotransformation 59
stirred tank bioreactor 206, 310
STM, see scanning tunnelling microscopy
stop codon 151
stop transfer sequence 222
strain 302
    development 303
    improvement 303
    isolation 54, 304
strain depositories, culture collections 96
strategic alliance 306
streptavidin 28, 58
streptokinase 61
    cost of 281
Streptomyces 213
    producing antibiotics 15
Strobel, G. 44
sub-cloning 278
substrate 2
substrate channelling 307
substrate specificity altered by protein
    engineering 266
sugar, sucrose
    and biofuels 38
    as biomass 46
    glucose isomerase 158
sugars and glycation 160
sugars in biology 160
sulphides in coal 105
sulphonyl urea herbicides 169
supercritical fluid, SCF
    enzymology 307
    organic phase catalysis 236
support for biotechnology 308
support material for affinity
    chromatography 4
surface acoustic wave, SAW 189
surface adherent cells 11
surface culture 210
surface plasmon resonance, SPR 234
suspended cell fermenter 221
SV40 T antigen 183
SwissProt 41
symbiotic bacteria for nitrogen fixation
    228
synchrotron radiation 265
synzymes 39

T-antigen, SV40 183
T-DNA 5
TAG, triacylglycerol 256
Tam, J. 326
tank bioreactors 310
Taq polymerase 241, 277
targetted drug delivery 311
tat gene 231

I_c, critical temperature 307
telomere 336
teratoma 79
tetraploid fish 23
texture modification by food processing
  using enzymes 137
thalassemia 152
thalidomide 84
thaumatin 23
Thaumatococcus danielli 23
thermal sensor 312
thermophile 312
Thermus acquaticus 241
thin layer chromatography, TLC 86
Thiobacillus 209
  Thiobacillus ferrooxidans 164
thrombin 4
thrombolytic 61, 264
thyrotropin-releasing hormone, TRH 244
Ti plasmid 5
  manipulated using recombination 172
TILs 192
tissue culture 313
  hairy cell root culture for plants 167
  micropropagation of plants 214
  to make monoclonal antibodies 221
tissue explant 253
tissue necrosis factor, TNF 98
  and gene therapy 147
  in immunotherapy 191
tissue plasminogen activator 61
tissue polypeptide antigen, tumour marker
  324
titre of antibody 185, 220
TLC, *see* thin layer chromatography
TNF, *see* tissue necrosis factor
tomato 20
totipotent 251
  cells to make chimera 79
  stem cells 300
tower fermenter 53
toxic chemical detection by immobilized cell
  biosensor 182
toxicity screening 335
toxins 314
  genes in transgenic animals 320
  proteins in immunotoxins 192
tPA, tissue plasminogen activator 61
  cost of 281
TPA, tissue polypeptide antigen, tumour
  marker 324
tracer 185
transdermal delivery 114
transduction 315
transesterification
  and lipases 200
  in organic phase catalysis 256
  of plant oils 236
transfection 315

of animals 317
and genoceuticals 155
of plants 254
transformation 316
  of cells to make cell line 74
  by electroporation 118
  of plants 255
transgenic animals 147, 153, 317
  applications 318
  made from chimera 79
  disease models 232, 319
  as expression system 133
  and homologous recombination 173
  made using retrovirus 284
transition phase of cell growth 71
transition state 66
transmissible encephalopathies 321
transomics 317
transplant technology 234
transposon 321
treatment protocol programme 322
trees manipulated using *Agrobacterium*
  tumefaciens 6
trehalose as stabilizer for freeze-drying
  138
TRH, *see* thyrotropin-releasing hormone
triacylglycerol 256
triazine dyes 38
triazine herbicides 169
trickling bed system of sewage treatment
  294
trienoic plant oils 256
triple DNA, triplex DNA 323
triplets 150
triploid fish 23
trout 22
  used as biosensor 55
tryptophan 10
  and eosinophilia–myalgia syndrome
    176
  hydrophobicity 10
tumour infiltrating lymphocytes 192
tumour marker 324
turbidostat 24
turbines for stirring 310
Turner's syndrome 173
two phase aqueous extraction 272
two phase separation 270
type culture collections 96
Type I, II, III fermentations 135

ultrafiltration 92
ultraviolet light
  in physical containment 302
  for sterilization 249
unilamellar liposomes 202
Unilever 298
Universal, tube 300

unsaturated fat 200
untranslated region 150
uranium extraction 164
urease 196

vaccines 326
  AIDS 13
  anti-idiotype antibody 8
  live vaccine 205
  vaccinia virus 327
vaccinia virus 327
vacuum blot 62
vapour phase storage in liquid nitrogen 95
variable region of antibody structure 18
variety, of plant 302
vector 328
  adenovirus 1
  baculovirus 26
  expression system 132
  for secretion 293
Venter, C. 68, 240
vertical integration 329
vinca alkaloids 292
viral insecticides 27
viral vaccine 205, 330
  and vaccinia virus 327
virtual reality 218
viruses
  as biocontrol agents 44
  used for cell fusion 227
  diseases 71
  as plasmids 259
vitamin C 59
vitamin K 262

walking 332
Waring blender 69
Warnock report 120
waste disposal 123
  in wood manufacture 334
water stress 238
weed control 44
western blot 63

whey as a fermentation substrate 127, 136
WHO, World Health Organisation
  culture collections 96
  microorganism safety classification 210
whooping cough 314
Winters, G. 99
wood 34, 333
wool 334
  from transgenic animals 319
wound care 159
wound debridement 264

X-ray crystallography of proteins 265
  and bioinformatics 274
  and rational drug design 41
xenobiotics 335
Xenova 15
Xoma 191
xylanase 333

YAC, yeast artificial chromosome 145, 335
  used for walking 332
yeast cloning vectors 336
yeast episomal plasmid, YEp 337
yeast extract as a fermentation substrate 136
yeast integrating plasmid, YIp 337
yeast 213
  genome project 156
  to make haemoglobin 60
YEp, see yeast episomal plasmid
YIp, see yeast integrating plasmid
Yuk factor 337
  and BST 64
  in bioethics 36

zinc finger motifs 222
zinc purified using microbial mining 209
zonal rotor 76
zone centrifugation 75
zooplankton 46

46   29